Problems Supplement to Accompany

Vector Mechanics for Engineers: Dynamics

Problems Supplement to Accompany

Vector Mechanics for Engineers

Dynamics

Fifth Edition

Ferdinand P. Beer
Lehigh University

E. Russell Johnston, Jr.
University of Connecticut

McGraw-Hill, Inc.
New York St. Louis San Francisco Auckland Bogotá Caracas
Hamburg Lisbon London Madrid Mexico Milan Montreal
New Delhi Oklahoma City Paris San Juan São Paulo Singapore
Sydney Tokyo Toronto

This book is printed on recycled paper containing a minimum of 50% total recycled fiber with 10% postconsumer de-inked fiber.

PROBLEMS SUPPLEMENT TO ACCOMPANY VECTOR MECHANICS FOR ENGINEERS: Dynamics

1 2 3 4 5 6 7 8 9 0 MAL MAL 9 0 9 8 7 6 5 4 3 2

ISBN 0-07-005009-0

The editor was John J. Corrigan;
the production supervisor was Phil Galea.
Malloy Lithographing, Inc., was printer and binder.

Contents

11 KINEMATICS OF PARTICLES 1
Sections 11.1 to 11.3 1
Sections 11.4 to 11.6 3
Sections 11.7 and 11.8 6
Sections 11.9 to 11.12 9
Section 11.13 and 11.14 12

12 KINETICS OF PARTICLES: NEWTON'S SECOND LAW 14
Sections 12.1 to 12.6 14
Sections 12.7 to 12.10 19
Sections 12.11 to 12.13 21

13 KINETICS OF PARTICLES: ENERGY AND MOMENTUM METHODS 24
Sections 13.1 to 13.5 24
Sections 13.6 to 13.9 27
Sections 13.10 and 13.11 31
Sections 13.12 to 13.15 33

14 SYSTEMS OF PARTICLES 36
Sections 14.1 to 14.6 36
Sections 14.7 to 14.9 39
Sections 14.10 to 14.12 41

15 KINEMATICS OF RIGID BODIES 46
Sections 15.1 to 15.4 46
Sections 15.5 and 15.6 49
Section 15.7 51
Sections 15.8 and 15.9 53
Sections 15.10 and 15.11 56
Sections 15.12 and 15.13 58
Sections 15.14 and 15.15 62

16 PLANE MOTION OF RIGID BODIES: FORCES AND ACCELERATIONS 65
Sections 16.1 to 16.7 65
Section 16.8 71

17 PLANE MOTION OF RIGID BODIES: ENERGY AND MOMENTUM METHODS 78
Sections 17.1 to 17.7 78
Sections 17.8 to 17.10 83
Sections 17.11 and 17.12 86

18 KINETICS OF RIGID BODIES IN THREE DIMENSIONS 90
Sections 18.1 to 18.4 90
Sections 18.5 to 18.8 94
Sections 18.9 to 18.11 99

19 MECHANICAL VIBRATIONS 101
Sections 19.1 to 19.3 101
Section 19.5 103
Section 19.6 106
Section 19.7 108
Sections 19.8 to 19.10 110

Answers 113

CHAPTER 11
KINEMATICS OF PARTICLES

Answers to all problems are given at the end of this booklet.

SECTIONS 11.1 to 11.3

11.1 The motion of a particle is defined by the relation $x = 2t^3 - 9t^2 + 12$, where x is expressed in inches and t in seconds. Determine the time, position, and acceleration when $v = 0$.

11.2 The motion of a particle is defined by the relation $s = 2t^3 - 15t^2 + 36t - 10$, where s is expressed in inches and t in seconds. Determine the position, velocity, and acceleration when $t = 4$ s.

11.3 The motion of a particle is defined by the relation $x = t^2 - 10t + 30$, where x is expressed in meters and t in seconds. Determine (*a*) when the velocity is zero, (*b*) the position and the total distance traveled when $t = 8$ s.

11.4 The motion of a particle is defined by the relation $x = \frac{1}{3}t^3 - 3t^2 + 8t + 2$, where x is expressed in meters and t in seconds. Determine (*a*) when the velocity is zero, (*b*) the position and the total distance traveled when the acceleration is zero.

11.5 The acceleration of a particle is directly proportional to the time t. At $t = 0$, the velocity of the particle is $v = -16$ m/s. Knowing that both the velocity and the position coordinate are zero when $t = 4$ s, write the equations of motion for the particle.

11.6 The acceleration of a particle is defined by the relation $a = -2\ \text{m/s}^2$. If $v = +8$ m/s and $x = 0$ when $t = 0$, determine the velocity, position, and total distance traveled when $t = 6$ s.

11.7 The acceleration of a particle is defined by the relation $a = kt^2$. (*a*) Knowing that $v = -250$ in./s when $t = 0$ and that $v = +250$ in./s when $t = 5$ s, determine the constant k. (*b*) Write the equations of motion knowing also that $x = 0$ when $t = 2$ s.

11.8 The acceleration of a particle is defined by the relation $a = 18 - 6t^2$. The particle starts at $t = 0$ with $v = 0$ and $x = 100$ in. Determine (*a*) the time when the velocity is again zero, (*b*) the position and velocity when $t = 4$ s, (*c*) the total distance traveled by the particle from $t = 0$ to $t = 4$ s.

11.9 The acceleration of a particle moving in a straight line is directed toward a fixed point O and is inversely proportional to the distance of the particle from O. At $t = 0$, the particle is 8 in. to the right of O, has a velocity of 16 in./s to the right, and has an acceleration of 12 in./s^2 to the left. Determine (*a*) the velocity of the particle when it is 12 in. away from O, (*b*) the position of the particle at which its velocity is zero.

11.10 The acceleration of a particle is defined by the relation $a = -kx^{-2}$. The particle starts with no initial velocity at $x = 12$ in., and it is observed that its velocity is 8 in./s when $x = 6$ in. Determine (*a*) the value of k, (*b*) the velocity of the particle when $x = 3$ in.

11.11 The acceleration of a particle is defined by the relation $a = 21 - 12x^2$, where a is expressed in m/s^2 and x in meters. The particle starts with no initial velocity at the position $x = 0$. Determine (*a*) the velocity when $x = 1.5$ m, (*b*) the position where the velocity is again zero, (*c*) the position where the velocity is maximum.

11.12 The acceleration of an oscillating particle is defined by the relation $a = -kx$. Find the value of k such that $v = 10$ m/s when $x = 0$ and $x = 2$ m when $v = 0$.

11.13 The acceleration of a particle is defined by the relation $a = -10v$, where a is expressed in m/s^2 and v in m/s. Knowing that at $t = 0$ the velocity is 30 m/s, determine (*a*) the distance the particle will travel before coming to rest, (*b*) the time required for the particle to come to rest, (*c*) the time required for the velocity of the particle to be reduced to 1 percent of its initial value.

11.14 The acceleration of a particle is defined by the relation $a = -0.0125v^2$, where a is the acceleration in m/s^2 and v is the velocity in m/s. If the particle is given an initial velocity v_0, find the distance it will travel (*a*) before its velocity drops to half the initial value, (*b*) before it comes to rest.

11.15 It has been determined experimentally that the magnitude in ft/s^2 of the deceleration due to air resistance of a projectile is $0.001v^2$, where v is expressed in ft/s. If the projectile is released from rest and keeps pointing downward, determine its velocity after it has fallen 500 ft. (*Hint.* The total acceleration is $g - 0.001v^2$, where $g = 32.2$ ft/s^2.)

11.16 The acceleration of a particle is defined by the relation $a = -0.004v^2$, where a is the acceleration in ft/sec^2 and v is the velocity in ft/sec. If the particle is given an initial velocity v_0, find the distance it will travel (*a*) before its velocity drops to half the initial value, (*b*) before it comes to rest.

SECTIONS 11.4 to 11.6

11.17 An automobile travels 240 m in 30 s while being accelerated at a constant rate of $0.2\ \mathrm{m/s^2}$. Determine (*a*) its initial velocity, (*b*) its final velocity, (*c*) the distance traveled during the first 10 s.

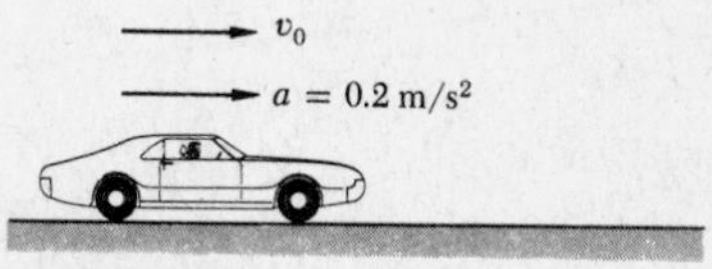

Fig. P11.17

11.18 A stone is released from an elevator moving up at a speed of 5 m/s and reaches the bottom of the shaft in 3 s. (*a*) How high was the elevator when the stone was released? (*b*) With what speed does the stone strike the bottom of the shaft?

11.19 A stone is thrown vertically upward from a point on a bridge located 135 ft above the water. Knowing that it strikes the water 4 s after release, determine (*a*) the speed with which the stone was thrown upward, (*b*) the speed with which the stone strikes the water.

11.20 A motorist is traveling at 45 mi/h when he observes that a traffic light 800 ft ahead of him turns red. The traffic light is timed to stay red for 15 s. If the motorist wishes to pass the light without stopping just as it turns green again, determine (*a*) the required uniform deceleration of the car, (*b*) the speed of the car as it passes the light.

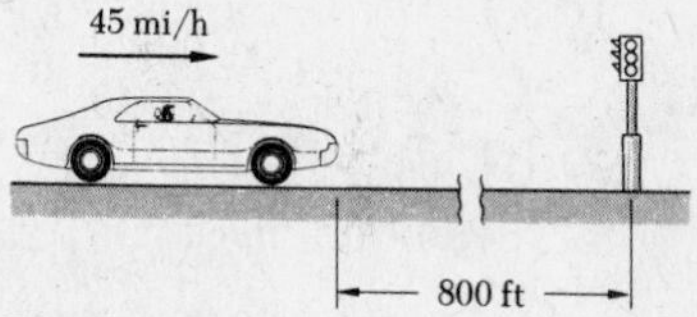

Fig. P11.20

11.21 A train starts at a station and accelerates uniformly at a rate of $0.6\ \mathrm{m/s^2}$ until it reaches a speed of 24 m/s; it then proceeds at a constant speed of 24 m/s. Determine the time and the distance traveled if its average velocity is (*a*) 16 m/s, (*b*) 22 m/s.

11.22 A man jumps from a 20-ft cliff with no initial velocity. (*a*) How long does it take him to reach the ground, and with what velocity does he hit the ground? (*b*) If this takes place on the moon, where $g = 5.31\ \mathrm{ft/s^2}$, what are the values obtained for the time and velocity? (*c*) If a motion picture is taken on the earth, but if the scene is supposed to take place on the moon, how many frames per second should be used so that the scene would appear realistic when projected at the standard speed of 24 frames per second?

11.23 Two automobiles *A* and *B* are traveling in the same direction in adjacent highway lanes. Automobile *B* is stopped when it is passed by *A*, which travels at a constant speed of 36 km/h. Two seconds later automobile *B* starts and accelerates at a constant rate of $1.5\ \mathrm{m/s^2}$. Determine (*a*) when and where *B* will overtake *A*, (*b*) the speed of *B* at that time.

11.24 Drops of water are observed to drip from a faucet at uniform intervals of time. As any drop B begins to fall freely, the preceding drop A has already fallen 0.3 m. Determine the distance drop A will have fallen by the time the distance between A and B will have increased to 0.9 m.

11.25 Automobile A starts from O and accelerates at the constant rate of 4 ft/s². A short time later it is passed by truck B which is traveling in the opposite direction at a constant speed of 45 ft/s. Knowing that truck B passes point O, 25 s after automobile A started from there, determine when and where the vehicles passed each other.

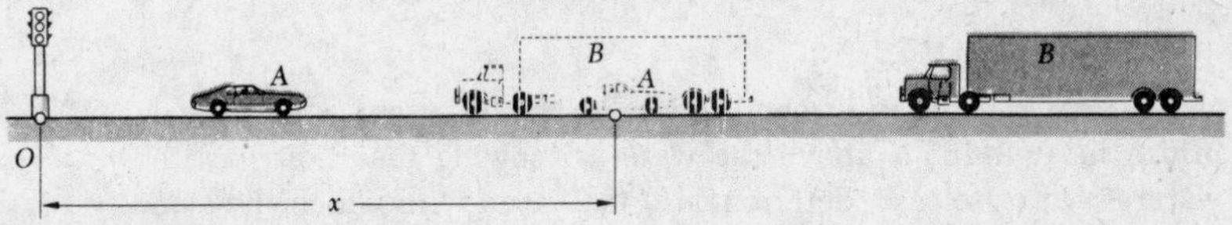

Fig. P11.25

11.26 An open-platform elevator is moving down a mine shaft at a constant velocity v_e when the elevator platform hits and dislodges a stone. Assuming that the stone starts falling with no initial velocity, (*a*) show that the stone will hit the platform with a relative velocity of magnitude v_e. (*b*) If $v_e = 16$ ft/s, determine when and where the stone will hit the elevator platform.

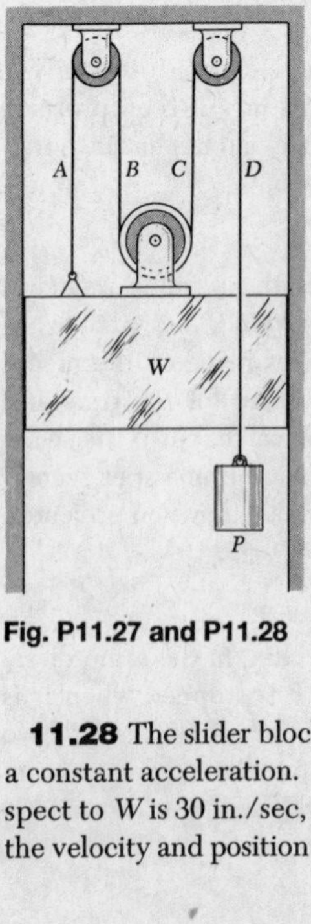

Fig. P11.27 and P11.28

11.27 The slider block W moves downward at a constant velocity of 12 in./sec. Find (*a*) the velocities of portions A, B, C, and D of the cable, (*b*) the relative velocity of P with respect to the weight W, (*c*) the relative velocity of portion B of the cable with respect to portion C.

11.28 The slider block W starts from rest and moves downward with a constant acceleration. If after 5 sec the relative velocity of P with respect to W is 30 in./sec, determine (*a*) the accelerations of W and P, (*b*) the velocity and position of W after 4 sec.

11.29 Knowing that block B moves downward with a constant velocity of 180 mm/s, determine (a) the velocity of block A, (b) the velocity of pulley D.

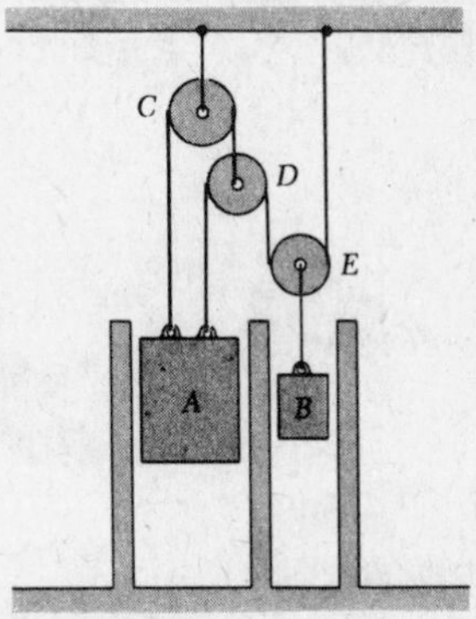

Fig. P11.29

11.30 The slider block B starts from rest and moves to the right with a constant acceleration. After 4 s the relative velocity of A with respect to B is 60 mm/s. Determine (a) the accelerations of A and B, (b) the velocity and position of B after 3 s.

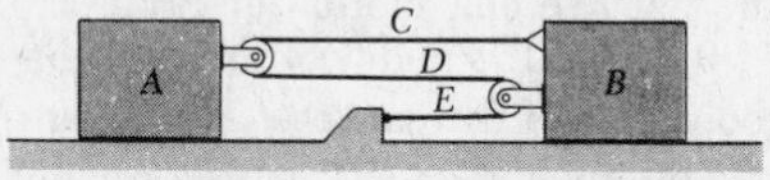

Fig. P11.30

11.31 Collars A and B start from rest and move with the following accelerations: $a_A = 75\ \text{mm/s}^2$ upward and $a_B = 150t\ \text{mm/s}^2$ downward. Determine (a) the time at which the velocity of block C is again zero, (b) the distance through which block C will have moved at that time.

11.32 (a) Choosing the positive sense upward, express the velocity of block C in terms of the velocities of collars A and B. (b) Knowing that both collars start from rest and move upward with the accelerations $a_A = 100\ \text{mm/s}^2$ and $a_B = 75\ \text{mm/s}^2$, determine the velocity of block C at $t = 4$ s and the distance through which it will have moved at that time.

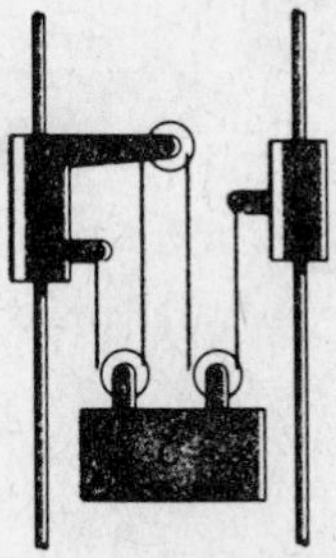
Fig. P11.31 and P11.32

11.33 The three blocks of Fig. 11.9 move with constant velocities. Find the velocity of each block, knowing that C is observed from B to move downward with a relative velocity of 9 ft/sec and A is observed from B to move downward with a relative velocity of 8 ft/sec.

11.34 Blocks A and C start from rest and move to the right with the following accelerations: $a_A = 12t\ \text{ft/s}^2$ and $a_C = 3\ \text{ft/s}^2$. Determine (a) the time at which the velocity of block B is zero, (b) the corresponding position of B.

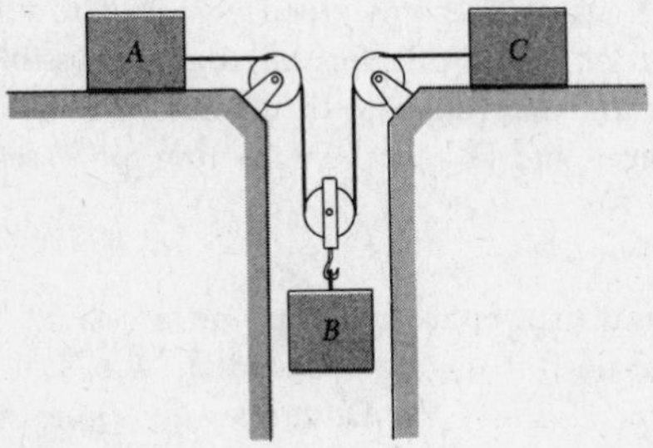

Fig. P11.34

SECTIONS 11.7 and 11.8

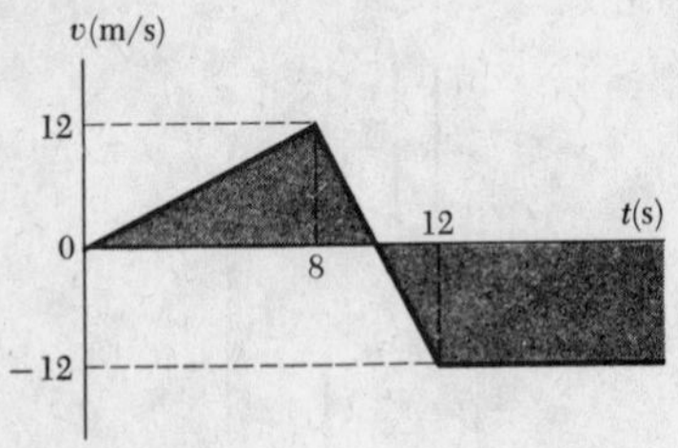

Fig. P11.35

11.35 A particle moves in a straight line with the velocity shown in the figure. Knowing that $x = -12$ m at $t = 0$, draw the a–t and x–t curves for $0 < t < 16$ s and determine (*a*) the total distance traveled by the particle after 12 s, (*b*) the two values of t for which the particle passes through the origin.

11.36 For the particle and motion of Prob. 11.35, plot the a–t and x–t curves for $0 < t < 16$ s and determine (*a*) the maximum value of the position coordinate of the particle, (*b*) the values of t for which the particle is at a distance of 15 m from the origin.

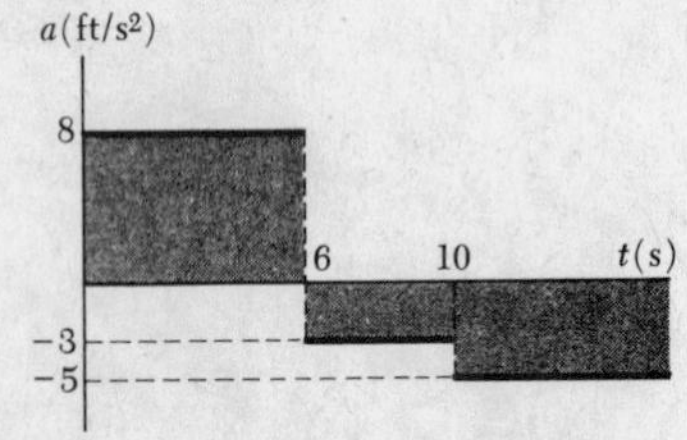

Fig. P11.37

11.37 A particle moves in a straight line with the acceleration shown in the figure. Knowing that it starts from the origin with $v_0 = -16$ ft/s, (*a*) plot the v–t and x–t curves for $0 < t < 16$ s, (*b*) determine its velocity, its position, and the total distance traveled after 12 s.

11.38 For the particle and motion of Prob. 11.37, plot the v–t and x–t curves for $0 < t < 16$ s and determine (*a*) the maximum value of the velocity of the particle, (*b*) the maximum value of its position coordinate.

11.39 A car starts from rest at point A and accelerates at the rate of 3 ft/sec^2 until it reaches a speed of 45 mph. It then proceeds at 45 mph until the brakes are applied; it comes to a stop at point B, 198 ft beyond the point where the brakes were applied. Knowing that the *average* speed of the car is 30 mph, determine (*a*) the time required for the car to travel from A to B, (*b*) the distance from A to B.

11 7 A motorist is traveling at 60 mi/h when he observes that a traffic signal 1000 ft ahead of him turns red. He knows that the signal is timed to stay red for 20 s. What should he do to pass the signal at 60 mi/h just as it turns green again? Draw the v–t curve, selecting the solution which calls for the smallest possible deceleration and acceleration, and determine (*a*) the common value of the deceleration and acceleration in ft/s^2, (*b*) the minimum speed reached in mi/h.

11.41 A train starts at a station and accelerates uniformly at a rate of 0.6 m/s^2 until it reaches a speed of 24 m/s; it then proceeds at the constant speed of 24 m/s. Determine the time and the distance traveled if its average velocity is (*a*) 16 m/s, (*b*) 22 m/s.

11.42 An express subway train and a train making local stops run on parallel tracks between stations A and E, which are 1400 m apart. The local train makes stops of 30-s duration at each of the stations B, C, and D; the express train proceeds to station E without any intermediate stop. Each train accelerates at a rate of $1.25\ \text{m/s}^2$ until it reaches a speed of 12.5 m/s; it then proceeds at that constant speed. As the train approaches its next stop, the brakes are applied, providing a constant deceleration of $1.5\ \text{m/s}^2$. If the express train leaves station A 4 min after the local train has left A, determine (*a*) which of the two trains will arrive at station E first, (*b*) how much later the other train will arrive at station E.

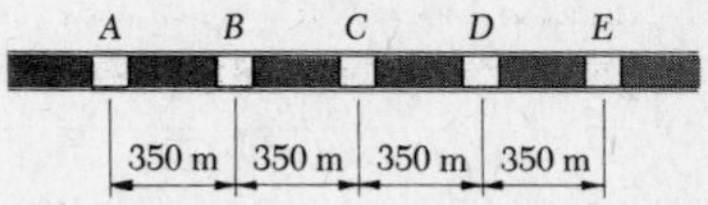

Fig. P11.42

11.43 A policeman on a motorcycle is escorting a motorcade which is traveling at 54 km/h. The policeman suddenly decides to take a new position in the motorcade, 70 m ahead. Assuming that he accelerates and decelerates at the rate of $2.5\ \text{m/s}^2$ and that he does not exceed at any time a speed of 72 km/h, draw the a–t and v–t curves for his motion and determine (*a*) the shortest time in which he can occupy his new position in the motorcade, (*b*) the distance he will travel in that time.

11.44 Two cars are traveling toward each other on a single-lane road at 16 and 12 m/s, respectively. When 120 m apart, both drivers realize the situation and apply their brakes. They succeed in stopping simultaneously, and just short of colliding. Assuming a constant deceleration for each car, determine (*a*) the time required for the cars to stop, (*b*) the deceleration of each car, and (*c*) the distance traveled by each car while slowing down.

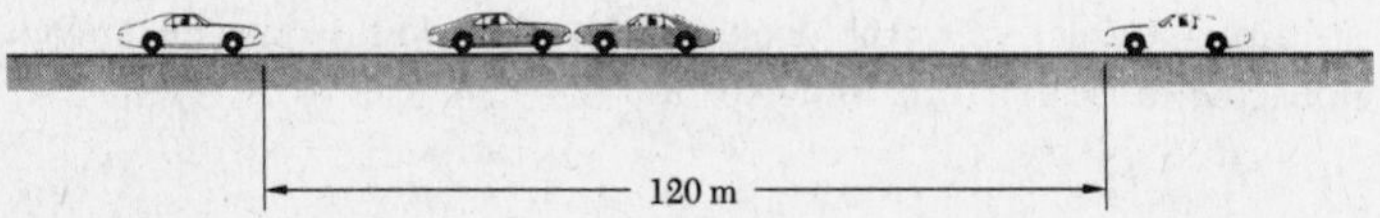

Fig. P11.44

11.45 A fighter plane flying horizontally in a straight line at 800 ft/sec is overtaking a bomber flying in the same straight line at 600 ft/sec. The pilot of the fighter plane fires an air-to-air missile at the bomber when his plane is 1,900 ft behind the bomber. The missile accelerates at a constant rate of $1{,}000\ \text{ft/sec}^2$ for 1 sec and then travels at a constant speed. (*a*) How many seconds after firing will the missile reach the bomber? (*b*) If both planes continue at constant speeds, what will be the distance between the planes when the missile strikes the bomber?

11.46 A police officer observes a car approaching at the unlawful speed of 60 mph. He gets on his motorcycle and starts chasing the car, just as it passes in front of him. After accelerating for 10 sec at a constant rate, the officer reaches his top speed of 75 mph. How long does it take him to overtake the car from the time he started? Draw the v–t and s–t curves for the car and the motorcycle.

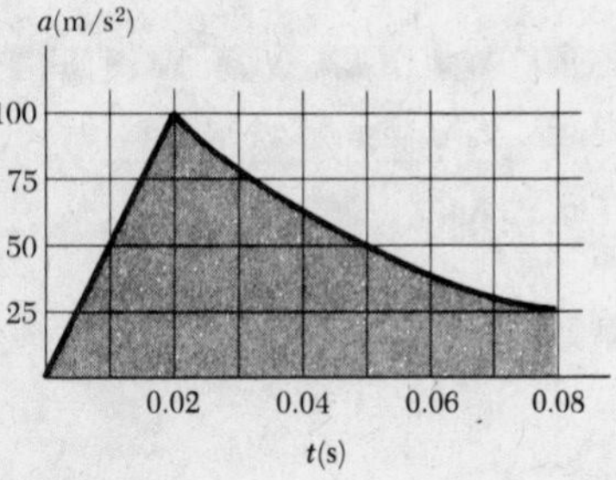

Fig. P11.47

11.47 The a–t curve shown was obtained during the motion of a test sled. Knowing that the sled started from rest at $t = 0$, determine the velocity and position of the sled at $t = 0.08$ s.

11.48 The acceleration record shown was obtained for a truck traveling on a straight highway. Knowing that the initial velocity of the truck was 18 km/h, determine the velocity and distance traveled when (*a*) $t = 4$ s, (*b*) $t = 6$ s.

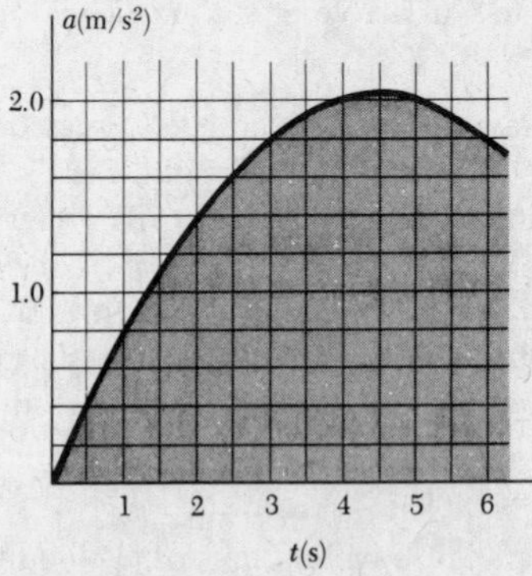

Fig. P11.48

11.49 A training airplane lands on an aircraft carrier and is brought to rest in 4 s by the arresting gear of the carrier. An accelerometer attached to the airplane provides the acceleration record shown. Determine by approximate means (*a*) the initial velocity of the airplane relative to the deck, (*b*) the distance the airplane travels along the deck before coming to rest.

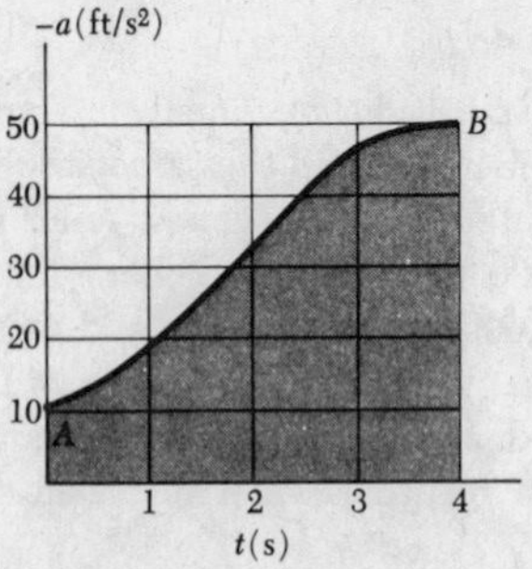

Fig. P11.49

11.50 Using the method of Sec. 11.8, obtain an approximate solution for Prob. 11.49, assuming that the a–t curve is a straight line from point A to point B.

SECTIONS 11.9 to 11.12

11.51 The motion of a particle is defined by the equations $x = \frac{1}{3}t^3 - 2t^2$ and $y = \frac{1}{2}t^2 - 2t$, where x and y are expressed in meters and t in seconds. Determine the velocity and acceleration when (a) $t = 1$ s, (b) $t = 3$ s.

11.52 In Prob. 11.51, determine (a) the time at which the value of the y coordinate is minimum, (b) the corresponding velocity and acceleration of the particle.

11.53 The motion of a particle is defined by the equations $x = e^{t/2}$ and $y = e^{-t/2}$, where x and y are expressed in feet and t in seconds. Show that the path of the particle is a rectangular hyperbola and determine the velocity and acceleration when (a) $t = 0$, (b) $t = 1$ s.

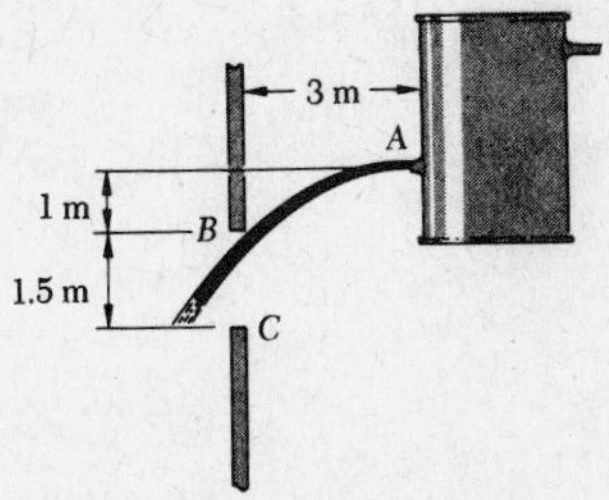

Fig. P11.55

11.54 The motion of a particle is defined by the equations $x = 5(1 - e^{-t})$ and $y = 5t/(t + 1)$, where x and y are expressed in feet and t in seconds. Determine the velocity and acceleration when $t = 1$ s.

11.55 Water issues at A from a pressure tank with a horizontal velocity $\mathbf{v}_0$. For what range of values v_0 will the water enter the opening BC?

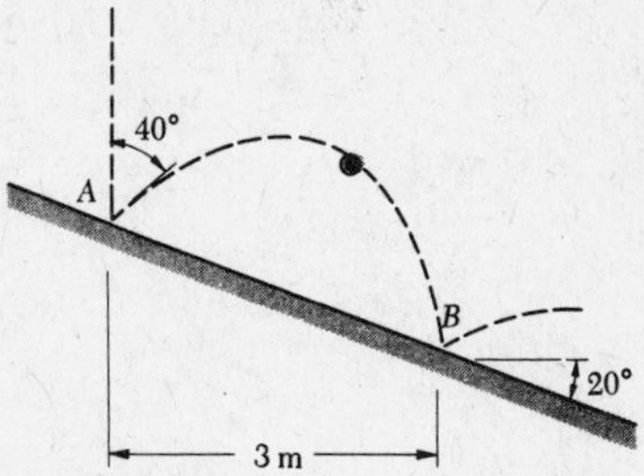

Fig. P11.56

11.56 A ball is dropped vertically onto a 20° incline at A; the direction of rebound forms an angle of 40° with the vertical. Knowing that the ball next strikes the incline at B, determine (a) the velocity of rebound at A, (b) the time required for the ball to travel from A to B.

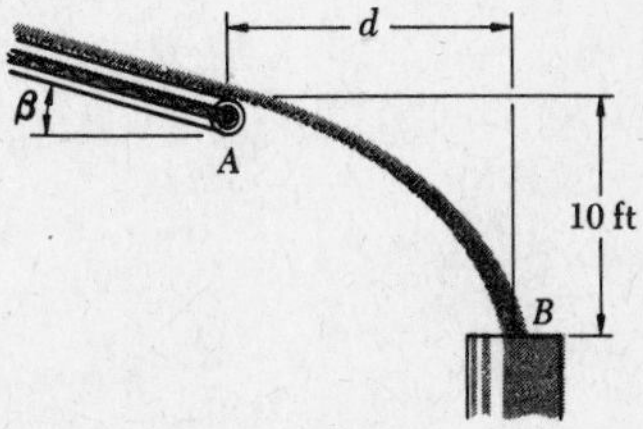

Fig. P11.57 and P11.58

11.57 Sand is discharged at A from a conveyor belt and falls into a collection pipe at B. Knowing that the conveyor belt forms an angle $\beta = 15°$ with the horizontal and moves at a constant speed of 20 ft/s, determine what the distance d should be so that the sand will hit the center of the pipe.

11.58 The conveyor belt moves at a constant speed of 12 ft/s. Knowing that $d = 8$ ft, determine the angle β for which the sand reaches the center of the pipe B.

11.59 If the maximum horizontal range of a given gun is R, determine the firing angle which should be used to hit a target located at a distance $\frac{1}{2}R$ on the same level.

11.60 Standing on the side of a hill, an archer shoots an arrow with an initial velocity of 250 ft/s at an angle $\alpha = 15°$ with the horizontal. Determine the horizontal distance d traveled by the arrow before it strikes the ground at B.

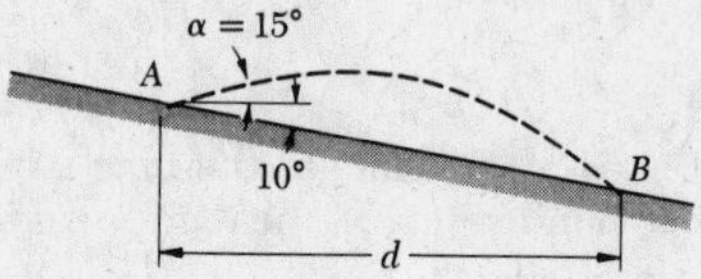

Fig. P11.60

11.61 As observed from a ship moving due east at 8 mph, the wind appears to blow from the southwest at 12 mph. Determine the magnitude and direction of the true wind velocity.

11.62 Two airplanes A and B are each flying at a constant altitude of 3000 ft. Plane A is flying due east at a constant speed of 300 mph while plane B is flying southwest at a constant speed of 450 mph. Determine the change in position of plane B relative to plane A which takes place during a 2-min interval.

11.63 An automobile and a train travel at the constant speeds shown. Three seconds after the train passes under the highway bridge the automobile crosses the bridge. Determine (*a*) the velocity of the train relative to the automobile, (*b*) the change in position of the train relative to the automobile during a 4-s interval, (*c*) the distance between the train and the automobile 5 s after the automobile has crossed the bridge.

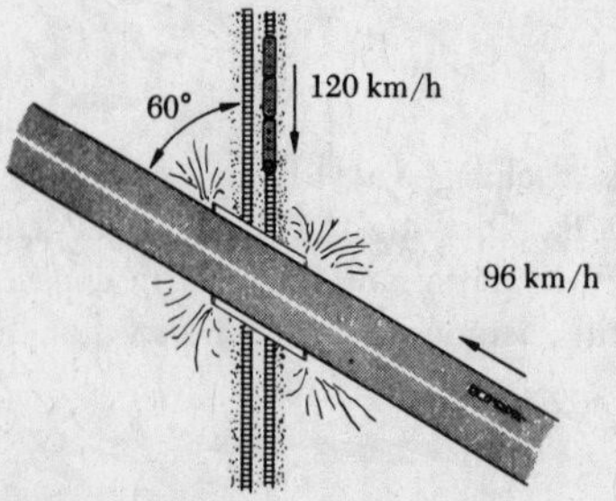

Fig. P11.63

11.64 As he passes a pole, a man riding in a truck tries to hit the pole by throwing a stone with a horizontal velocity of 20 m/s relative to the truck. Knowing that the speed of the truck is 40 km/h, determine (*a*) the direction in which he must throw the stone, (*b*) the horizontal velocity of the stone with respect to the ground.

11.65 An antiaircraft gun fires a shell as a plane passes directly over the position of the gun at an altitude of 6,000 ft. The muzzle velocity of the shell is 1,500 ft/sec. Knowing that the plane is flying horizontally at 450 mph, determine (*a*) the required firing angle if the shell is to hit the plane, (*b*) the velocity and acceleration of the shell relative to the plane at the time of impact.

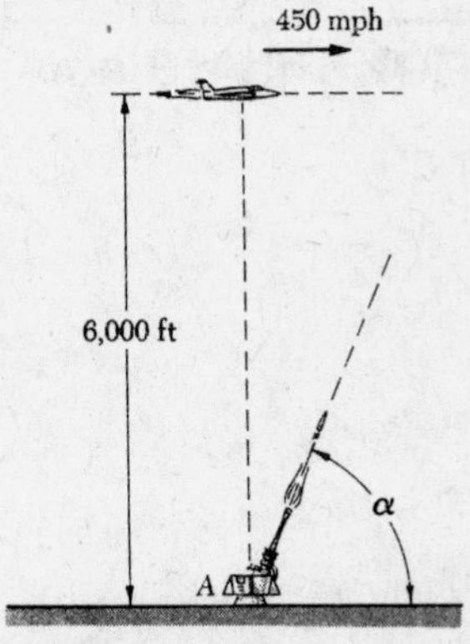

Fig. P11.65

11.66 Ship B is proceeding due north on the course shown. The shore gun at A is trained due east and will be fired in an attempt to hit the ship at point C. Knowing that the muzzle velocity of the gun is 2400 ft/s, determine (*a*) the required firing angle α, (*b*) the required angle β between AC and the line of sight at the time of firing.

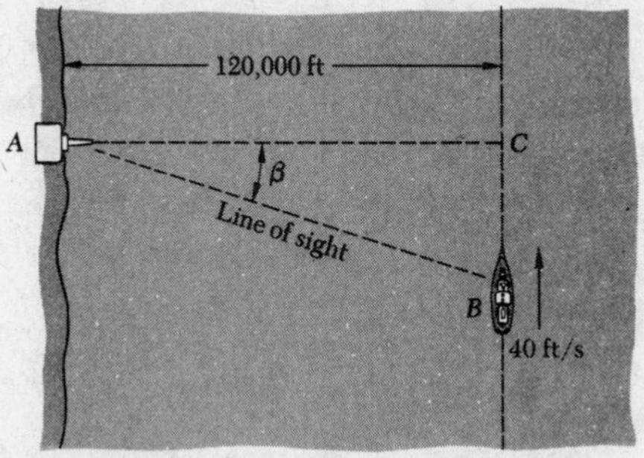

Fig. P11.66

11.67 An airplane is flying horizontally at an altitude of 2500 m and at a constant speed of 900 km/h on a path which passes directly over an antiaircraft gun. The gun fires a shell with a muzzle velocity of 500 m/s and hits the airplane. Knowing that the firing angle of the gun is 60°, determine the velocity and acceleration of the shell relative to the airplane at the time of impact.

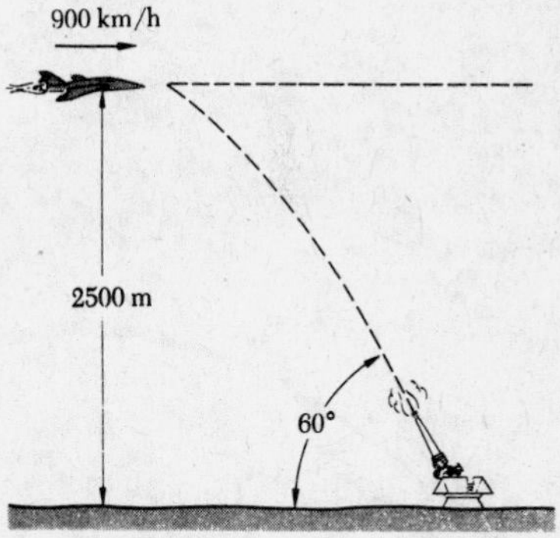

Fig. P11.67

11.68 Water is discharged at A with an initial velocity of 10 m/s and strikes a series of vanes at B. Knowing that the vanes move downward with a constant speed of 3 m/s, determine the velocity and acceleration of the water relative to the vane at B.

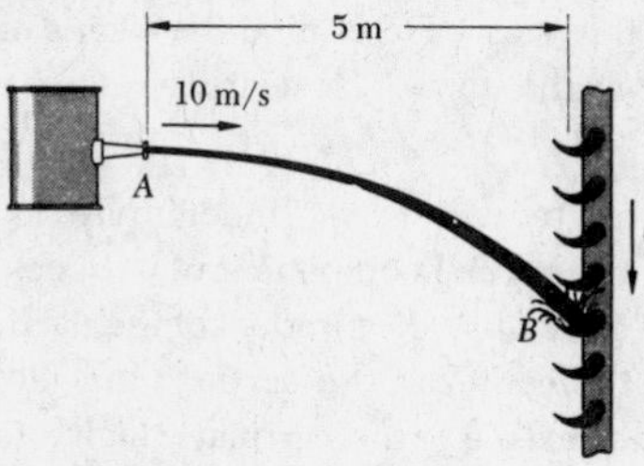

Fig. P11.68

11.69 A motorist starts from rest on a curve of 120-m radius and accelerates at the uniform rate of $0.9\ m/s^2$. Determine the distance that his automobile will travel before the magnitude of its total acceleration is $1.8\ m/s^2$.

11.70 A monorail train is traveling at a speed of 144 km/h along a curve of 1000-m radius. Determine the maximum rate at which the speed may be decreased if the total acceleration of the train is not to exceed $2\ m/s^2$.

11.71 A passenger train is traveling at 60 mi/h along a curve of 3000-ft radius. If the maximum total acceleration is not to exceed $g/10$, determine the maximum rate at which the speed may be decreased.

11.72 A motorist enters a curve of 500-ft radius at a speed of 45 mi/h. As he applies his brakes, he decreases his speed at a constant rate of $5\ ft/s^2$. Determine the magnitude of the total acceleration of the automobile when its speed is 40 mi/h.

11.73 In Prob. 11.60, determine the radius of curvature of the trajectory (*a*) immediately after the arrow has been shot, (*b*) as the arrow passes through its point of maximum elevation.

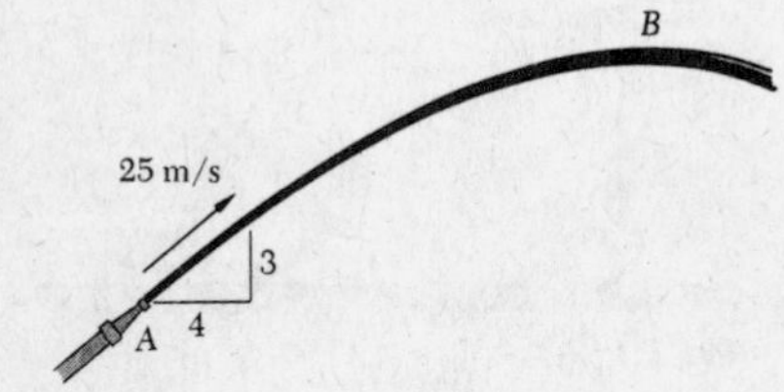

Fig. P11.74

11.74 A nozzle discharges a stream of water in the direction shown with an initial velocity of 25 m/s. Determine the radius of curvature of the stream (*a*) as it leaves the nozzle, (*b*) at the maximum height of the stream.

11.75 For each of the two firing angles obtained in Sample Prob. 11.8, determine the radius of curvature of the trajectory described by the projectile as it leaves the gun.

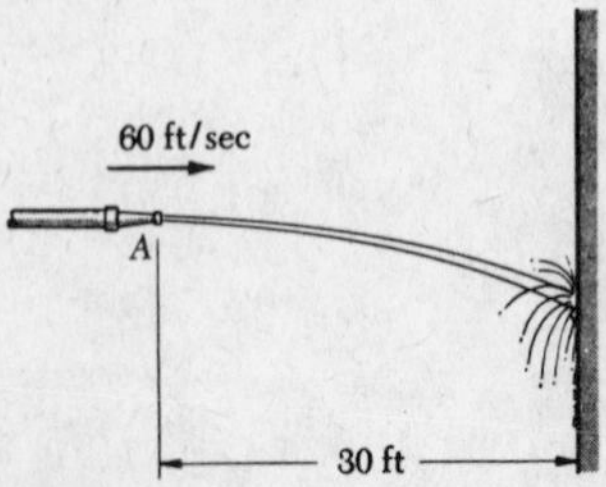

Fig. P11.76

11.76 A nozzle discharges a stream of water horizontally with an initial velocity of 60 ft/sec. Determine the radius of curvature of the stream (*a*) as it leaves the nozzle, (*b*) as it strikes the wall at *B*.

11.77 A satellite will travel indefinitely in a circular orbit around the earth if the normal component of its acceleration is equal to $g(R/r)^2$, where $g = 9.81\ m/s^2$, R = radius of the earth = 6370 km, and r = distance from the center of the earth to the satellite. Determine the height above the surface of the earth at which a satellite will travel indefinitely around the earth at a speed of 24×10^3 km/h.

11.78 Determine the speed of an earth satellite traveling in a circular orbit 480 km above the surface of the earth. (See information given in Prob. 11.77.)

11.79 Assuming the orbit of the moon to be a circle of radius 239,000 mi, determine the speed of the moon relative to the earth. (See information given in Prob. 11.77 and use $g = 32.2$ ft/s² and $R = 3960$ mi.)

11.80 The two-dimensional motion of a particle is defined by the relations $r = 60t^2 - 20t^3$ and $\theta = 2t^2$, where r is expressed in millimeters, t in seconds, and θ in radians. Determine the velocity and acceleration of the particle when (*a*) $t = 0$, (*b*) $t = 1$ s.

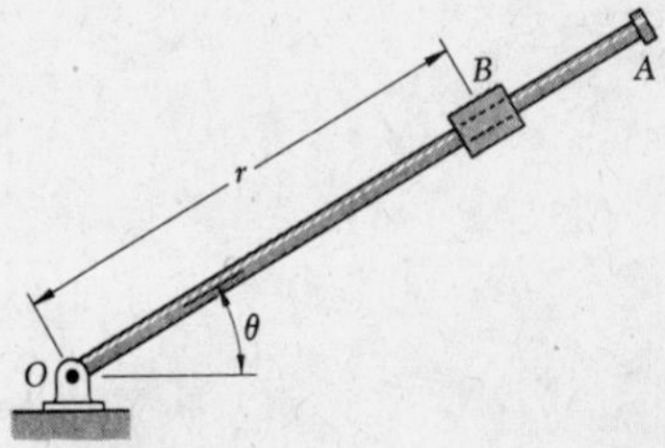

Fig. P11.80

11.81 The particle of Prob. 11.80 is at the origin at $t = 0$. Determine its velocity and acceleration as it returns to the origin.

11.82 As circle B rolls on the fixed circle A, point P describes a cardioid defined by the relations $r = 2b(1 + \cos 2\pi t)$ and $\theta = 2\pi t$. Determine the velocity and acceleration of P when (*a*) $t = 0.25$, (*b*) $t = 0.50$.

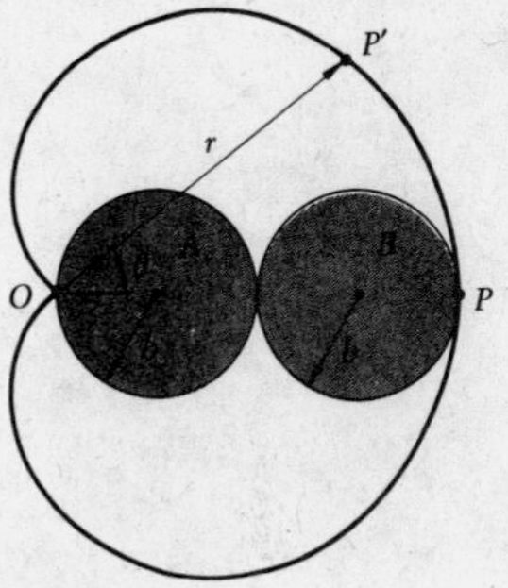

Fig. P11.82

11.83 The two-dimensional motion of a particle is defined by the relations $r = b \sin \pi t$ and $\theta = \frac{1}{2}\pi t$. Determine the velocity and acceleration of the particle when (*a*) $t = \frac{1}{2}$, (*b*) $t = 1$.

11.84 The pin at B is free to slide along the circular slot and along the rotating rod OC. If pin B slides counterclockwise around the circular slot at a constant speed v_0, determine the rate $d\theta/dt$ at which rod OC rotates and the radial component v_r of the velocity of the pin B (*a*) when $\phi = 0°$, (*b*) when $\phi = 90°$.

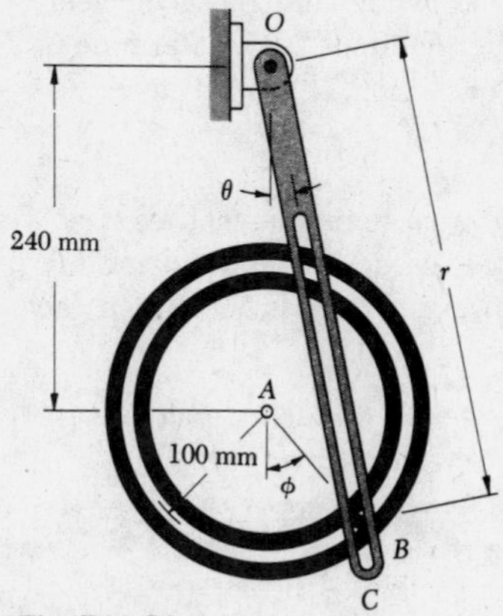

Fig. P11.84

CHAPTER 12
KINETICS OF PARTICLES: NEWTON'S SECOND LAW

SECTIONS 12.1 to 12.6

12.1 The 3-kg collar was moving down the rod with a velocity of 3 m/s when a force **P** was applied to the horizontal cable. Assuming negligible friction between the collar and the rod, determine the magnitude of the force **P** if the collar stopped after moving 1 m more down the rod.

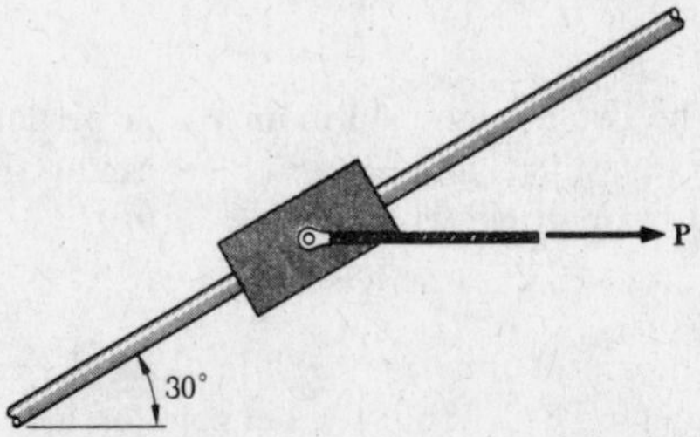

Fig. P12.1

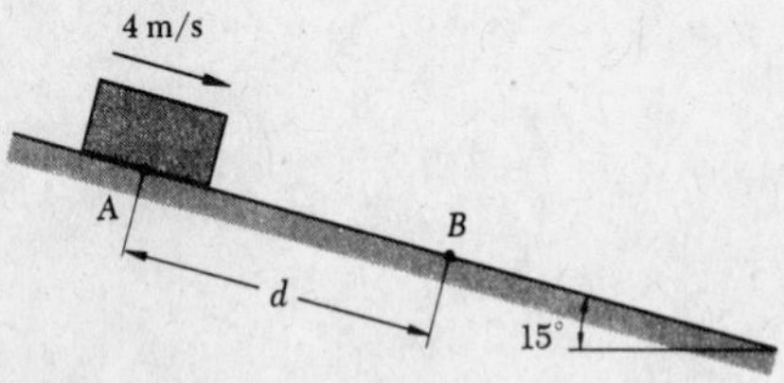

Fig. P12.2

12.2 A 5-kg package is projected down the incline with an initial velocity of 4 m/s. Knowing that the coefficient of friction between the package and the incline is 0.35, determine (*a*) the velocity of the package after 3 m of motion, (*b*) the distance *d* at which the package comes to rest.

12.3 The subway train shown travels at a speed of 30 mi/h. Determine the force in each coupling when the brakes are applied, knowing that the braking force is 5000 lb on each car.

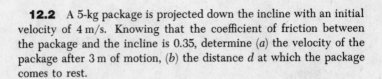

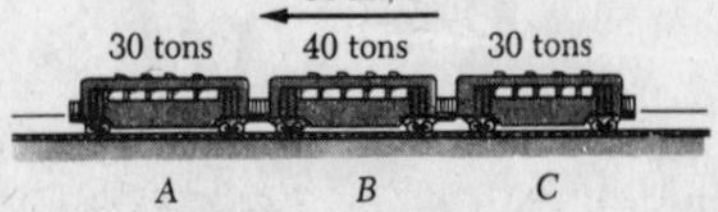

Fig. P12.3

12.4 Two packages are placed on an incline as shown. The coefficient of friction is 0.25 between the incline and package A, and 0.15 between the incline and package B. Knowing that the packages are in contact when released, determine (a) the acceleration of each package, (b) the force exerted by package A on package B.

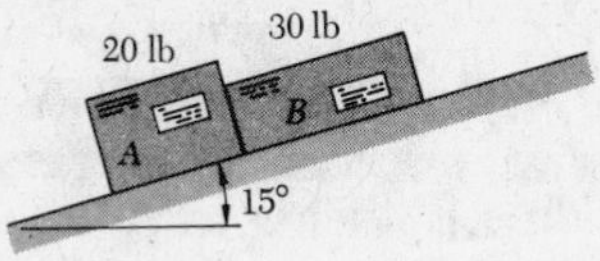

Fig. P12.4

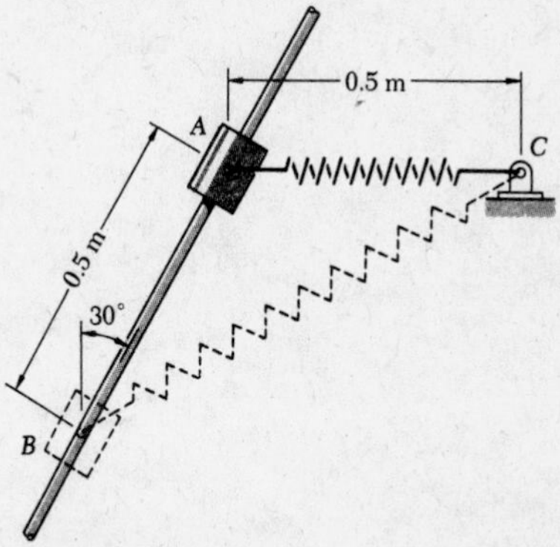

Fig. P12.5

12.5 A 4-kg collar slides without friction along a rod which forms an angle of 30° with the vertical. The spring, of constant $k = 150$ N/m, is unstretched when the collar is at A. Determine the initial acceleration of the collar if it is released from rest at point B.

12.6 The 100-kg block A is connected to a 25-kg counterweight B by the cable arrangement shown. If the system is released from rest, determine (a) the tension in the cable, (b) the velocity of B after 3 s, (c) the velocity of A after it has moved 1.2 m.

12.7 Block A is observed to move with an acceleration of 0.9 m/s^2 directed upward. Determine (a) the mass of block B, (b) the corresponding tension in the cable.

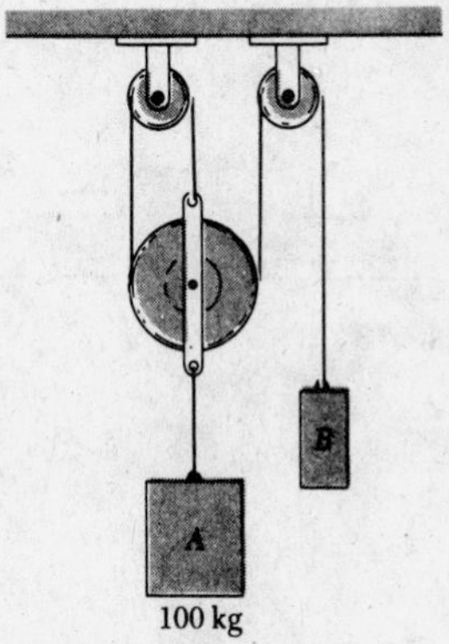

Fig. P12.6 and P12.7

12.8 The system shown is initially at rest. Neglecting the effect of friction, determine (a) the force **P** required if the velocity of collar B is to be 12 ft/s after it has moved 18 in. to the right, (b) the corresponding tension in the cable.

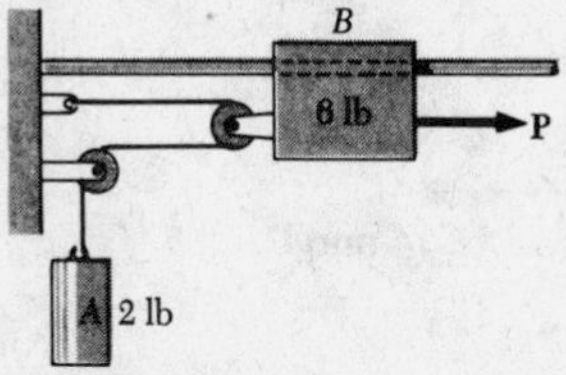

Fig. P12.8 and P12.9

12.9 A force **P** of magnitude 15 lb is applied to collar B, which is observed to move 3 ft in 0.5 s after starting from rest. Neglecting the effect of friction in the pulleys, determine the friction force that the rod exerts on collar B.

12.10 Knowing that the coefficient of friction is 0.30 at all surfaces of contact, determine (a) the acceleration of plate A, (b) the tension in the cable. (Neglect bearing friction in the pulley.)

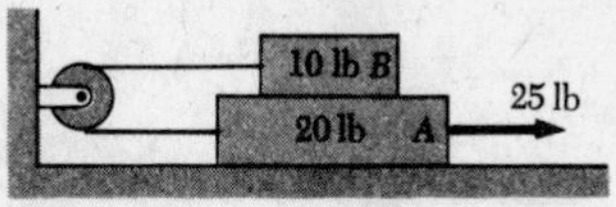

Fig. P12.10

12.11 A 30-kg crate rests on a 20-kg cart; the coefficient of static friction between the crate and the cart is 0.25. If the crate is not to slip with respect to the cart, determine (*a*) the maximum allowable magnitude of **P**, (*b*) the corresponding acceleration of the cart.

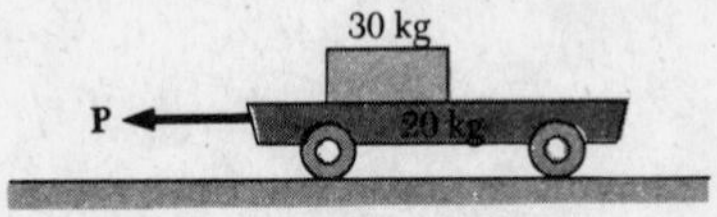

Fig. P12.11 and P12.12

12.12 The coefficients of friction between the 30-kg crate and the 20-kg cart are $\mu_s = 0.25$ and $\mu_k = 0.20$. If a force **P** of magnitude 150 N is applied to the cart, determine the acceleration (*a*) of the cart, (*b*) of the crate, (*c*) of the crate with respect to the cart.

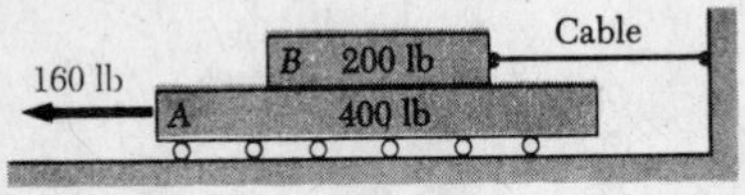

Fig. P12.13

12.13 The 400-lb plate *A* rests on small rollers which may be considered frictionless; the coefficient of friction between the two plates is $\mu = 0.20$. Determine the acceleration of plate *A* when a 160-lb force is suddenly applied to it, (*a*) for the system shown, (*b*) if the cable is cut.

12.14 Knowing that $\mu = 0.30$, determine the acceleration of each block when $m_A = m_B = m_C$.

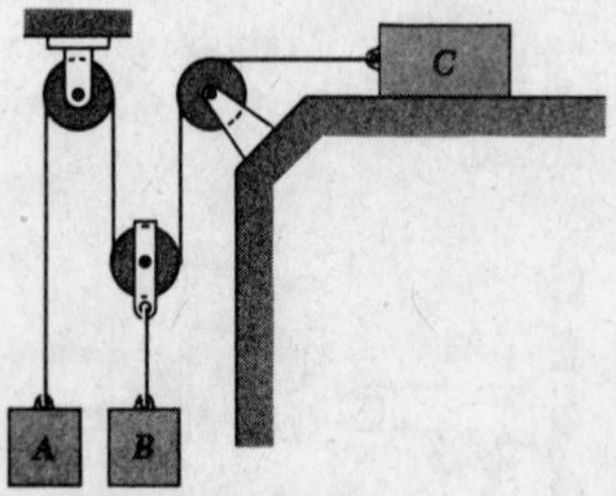

Fig. P12.14 and P12.15

12.15 Knowing that $\mu = 0.50$, determine the acceleration of each block when $m_A = 5$ kg, $m_B = 20$ kg, and $m_C = 15$ kg.

12.16 Knowing that blocks *B* and *C* strike the ground simultaneously and exactly 1 s after the system is released from rest, determine W_B and W_C in terms of W_A.

12.17 Determine the acceleration of each block when $W_A = 10$ lb, $W_B = 30$ lb, and $W_C = 20$ lb. Which block strikes the ground first?

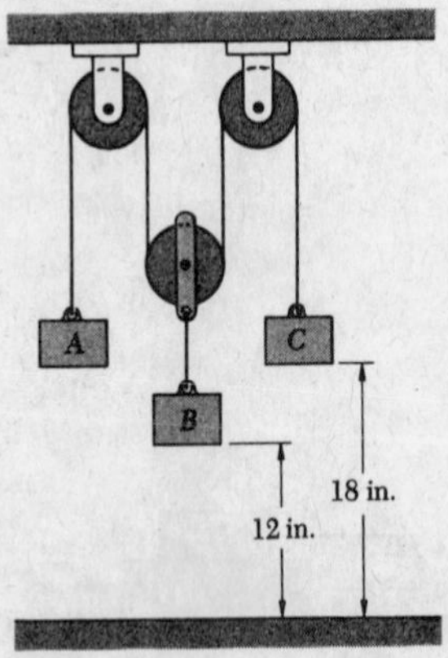

Fig. P12.16 and P12.17

12.18 A 3-kg ball is swung in a vertical circle at the end of a cord of length $l = 0.8$ m. Knowing that when $\theta = 60°$ the tension in the cord is 25 N, determine the instantaneous velocity and acceleration of the ball.

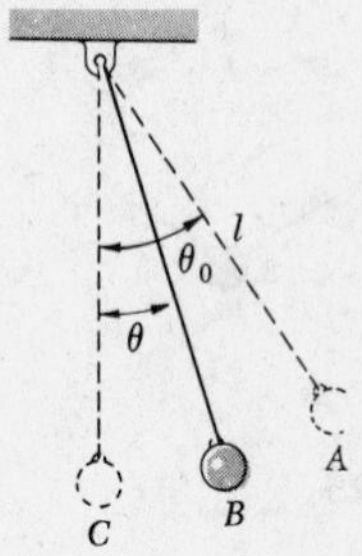

Fig. P12.18

12.19 Two wires AC and BC are each tied to a sphere at C. The sphere is made to revolve in a horizontal circle at a constant speed v. Determine the range of values of the speed v for which *both* wires are taut.

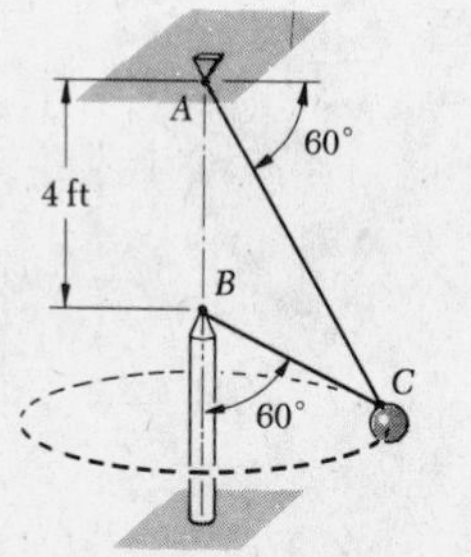

Fig. P12.19 and P12.20

12.20 Two wires AC and BC are each tied to a 10-lb sphere. The sphere is made to revolve in a horizontal circle at a constant speed v. Determine (a) the speed for which the tension is the same in both wires, (b) the corresponding tension.

12.21 A small ball of mass $m = 5$ kg is attached to a cord of length $L = 2$ m and is made to revolve in a horizontal circle at a constant speed v_0. Knowing that the cord forms an angle $\theta = 40°$ with the vertical, determine (a) the tension in the cord, (b) the speed v_0 of the ball.

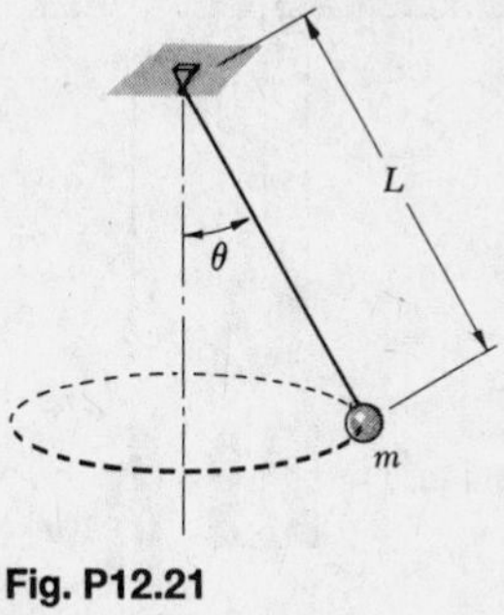

Fig. P12.21

12.22 Three automobiles are proceeding at a speed of 50 mi/h along the road shown. Knowing that the coefficient of friction between the tires and the road is 0.60, determine the tangential deceleration of each automobile if its brakes are suddenly applied and the wheels skid.

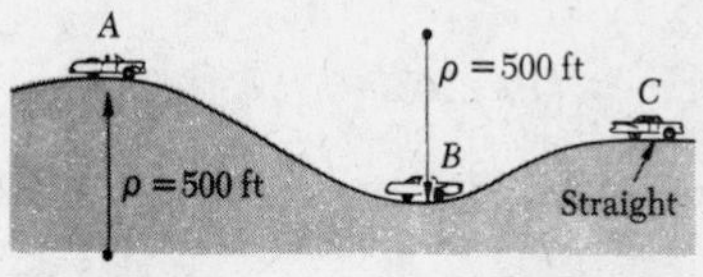

Fig. P12.22

12.23 A man swings a bucket full of water in a vertical plane in a circle of radius 0.75 m. What is the smallest velocity that the bucket should have at the top of the circle if no water is to be spilled?

12.24 The rod OAB rotates in a vertical plane at a constant rate such that the speed of collar C is 1.5 m/s. The collar is free to slide on the rod between two stops A and B. Knowing that the distance between the stops is only slightly larger than the collar and neglecting the effect of friction, determine the range of values of θ for which the collar is in contact with stop A.

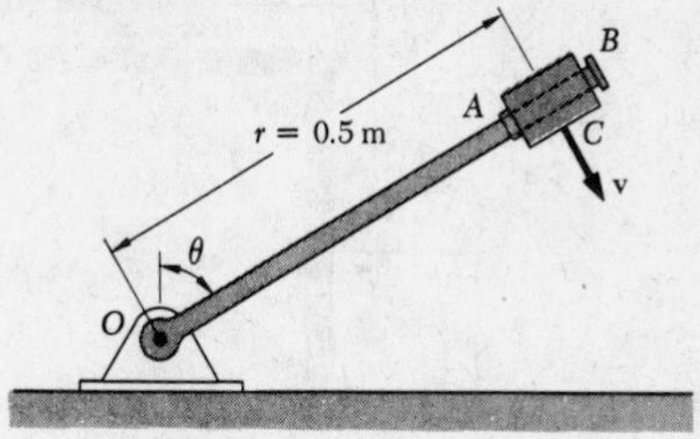

Fig. P12.24

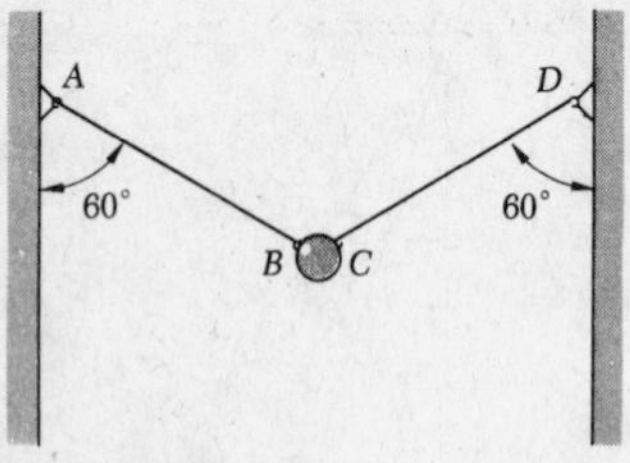

Fig. P12.25

12.25 A small sphere of weight W is held as shown by two wires AB and CD. Wire AB is then cut. Determine (*a*) the tension in wire CD before AB was cut, (*b*) the tension in wire CD and the acceleration of the sphere just after AB has been cut.

12.26 A small ball rolls at a speed v_0 along a horizontal circle inside the circular cone shown. Express the speed v_0 in terms of the height y of the path above the apex of the cone.

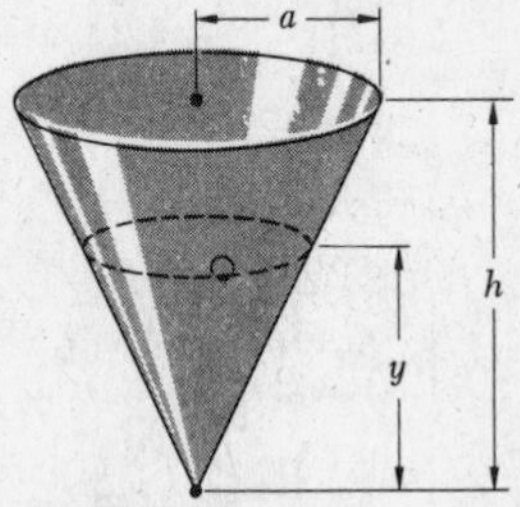

Fig. P12.26

12.27 A man on a motorcycle takes a turn on a flat unbanked road at 72 km/h. If the radius of the turn is 50 m, determine the minimum value of the coefficient of friction between the tires and the road which will ensure no skidding.

12.28 What angle of banking should be given to the road in Prob. 12.27 if the man on the motorcycle is to be able to take the turn at 72 km/h with a coefficient of friction $\mu = 0.30$?

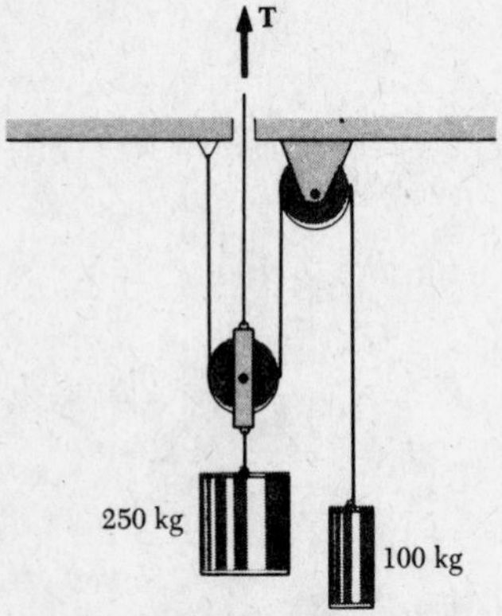

Fig. P12.29 and P12.30

12.29 Determine the required tension T if the acceleration of the 250-kg cylinder is to be (*a*) 2 m/s² upward, (*b*) 2 m/s² downward.

12.30 Determine the acceleration of the 100-kg cylinder if (*a*) $T = 1500$ N, (*b*) $T = 4000$ N.

12.31 Neglecting the effect of friction, determine (*a*) the acceleration of each block, (*b*) the tension in the cable.

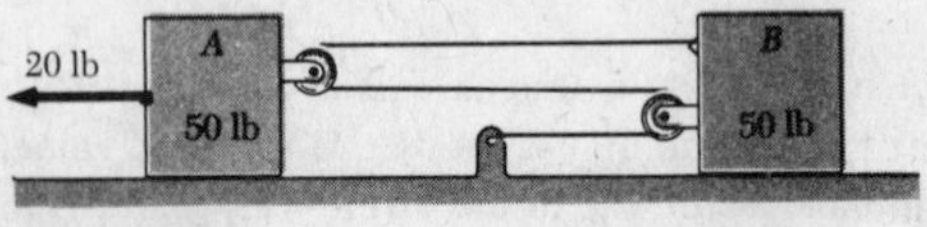

Fig. P12.31

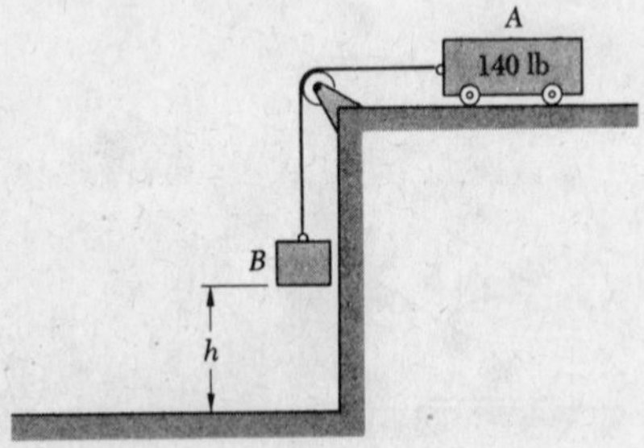

Fig. P12.32

12.32 The system shown is released from rest when $h = 4$ ft. (*a*) Determine the weight of block B, knowing that it strikes the ground with a speed of 9 ft/sec. (*b*) Attempt to solve part *a*, assuming the final speed to be 18 ft/sec; explain the difficulty encountered.

SECTIONS 12.7 to 12.10

12.33 The two-dimensional motion of particle B is defined by the relations $r = t^2 - \frac{1}{3}t^3$ and $\theta = 2t^2$, where r is expressed in meters, t in seconds, and θ in radians. If the particle has a mass of 2 kg and moves in a horizontal plane, determine the radial and transverse components of the force acting on the particle when (*a*) $t = 0$, (*b*) $t = 1$ s.

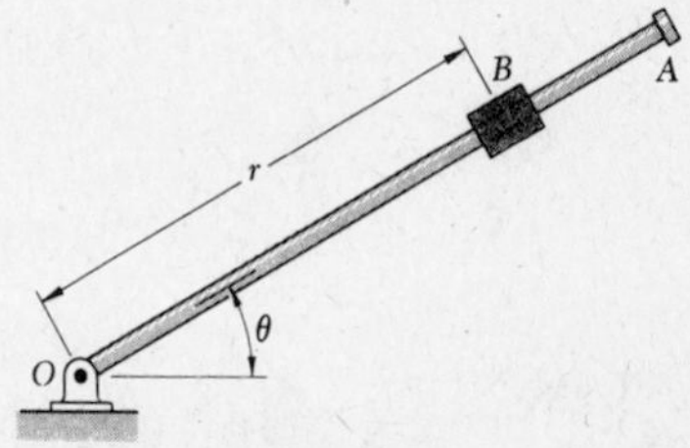

Fig. P12.33 and P12.34

12.34 For the motion defined in Prob. 12.33, determine the radial and transverse components of the force acting on the 2-kg particle as it returns to the origin at $t = 3$ s.

12.35 The two-dimensional motion of a particle is defined by the relations $r = 2 \sin \pi t$ and $\theta = \frac{1}{2}\pi t$, where r is expressed in feet, t in seconds, and θ in radians. If the particle weighs 3 lb and moves in a horizontal plane, determine the radial and transverse components of the force acting on the particle when (*a*) $t = \frac{1}{2}$ sec, (*b*) $t = 1$ sec.

12.36 A particle moves under a central force in a path defined by the equation $r = r_0/\cos n\theta$, where n is a positive constant. Using Eq. (12.27) show that the radial and transverse components of the velocity are $v_r = nv_0 \sin n\theta$ and $v_\theta = v_0 \cos n\theta$, where v_0 is the velocity of the particle for $\theta = 0$. What is the motion of the particle when $n = 0$ and when $n = 1$?

12.37 If a particle of mass m is attached to the end of a very light circular rod as shown in (*1*), the rod exerts on the mass a force **F** of magnitude $F = kr$ directed toward the origin O, as shown in (*2*). The path of the particle is observed to be an ellipse with semiaxes $a = 6$ in. and $b = 2$ in. (*a*) Knowing that the speed of the particle at A is 8 in./s, determine the speed at B. (*b*) Further knowing that the constant k/m is equal to 16 s^{-2}, determine the radius of curvature of the path at A and at B.

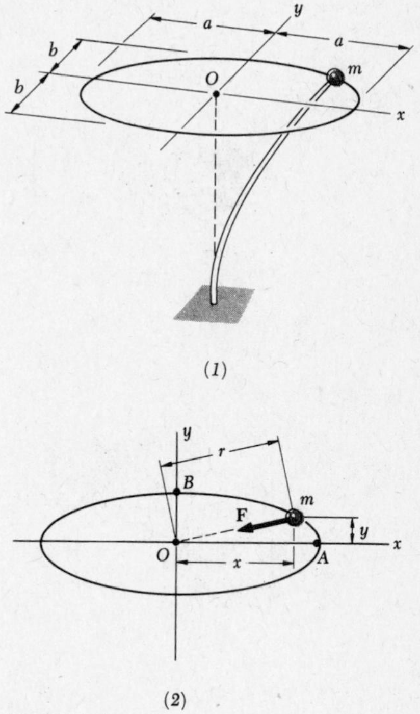

Fig. P12.37

12.38 Determine the mass of the earth from Newton's law of gravitation, knowing that it takes 94.14 min for a satellite to describe a circular orbit 300 mi above the surface of the earth.

12.39 Determine the time required for a spacecraft to describe a circular orbit 30 km above the moon's surface. It is known that the radius of the moon is 1740 km and that its mass is 81.3 times smaller than the mass of the earth.

12.40 Two solid steel spheres, each of radius 100 mm, are placed so that their surfaces are in contact. (*a*) Determine the force of gravitational attraction between the spheres, knowing that the density of steel is 7850 kg/m^3. (*b*) If the spheres are moved 2 mm apart and released with zero velocity, determine the approximate time required for their gravitational attraction to bring them back into contact. (*Hint.* Assume the gravitational forces to remain constant.)

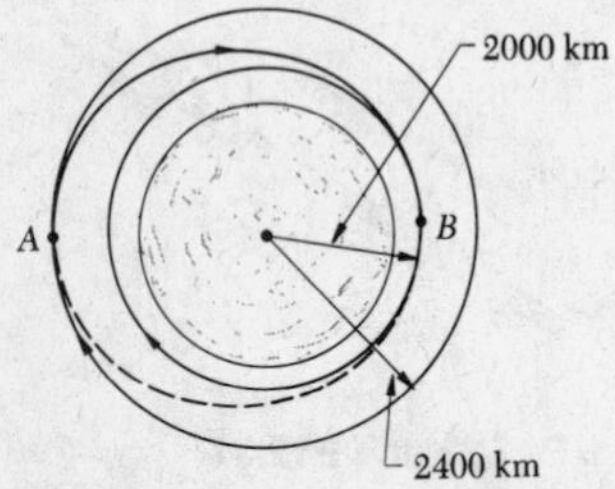

Fig. P12.41

12.41 An Apollo spacecraft describes a circular orbit of 2400-km radius around the moon with a velocity of 5140 km/h. In order to transfer it to a smaller circular orbit of 2000-km radius, the spacecraft is first placed on an elliptic path *AB* by reducing its velocity to 4900 km/h as it passes through *A*. Determine (*a*) the velocity of the spacecraft as it approaches *B* on the elliptic path, (*b*) the value to which its velocity must be reduced at *B* to insert it into the smaller circular orbit.

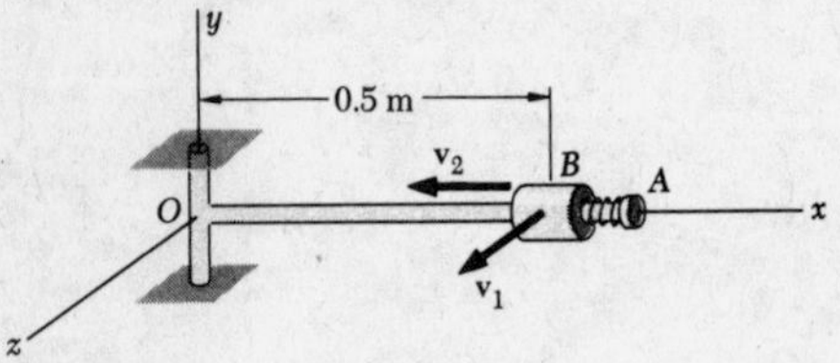

Fig. P12.42

12.42 Collar *B* may slide freely on rod *OA*, which in turn may rotate freely in the horizontal plane. The collar is describing a circle of radius 0.5 m with a speed $v_1 = 0.28$ m/s when a spring located between *A* and *B* is released, projecting the collar along the rod with an initial relative speed $v_2 = 0.96$ m/s. Neglecting the mass of the rod, determine the minimum distance between the collar and point *O* in the ensuing motion.

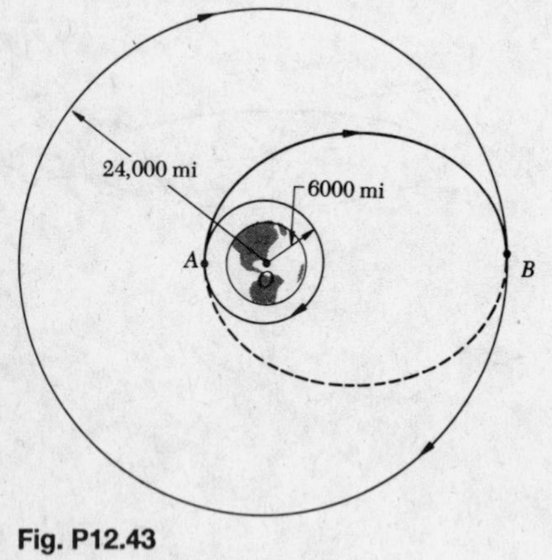

Fig. P12.43

12.43 A space tug describes a circular orbit of 6000-mi radius around the earth. In order to transfer it to a larger circular orbit of 24,000-mi radius, the tug is first placed on an elliptic path *AB* by firing its engine as it passes through *A*, thus increasing its velocity by 3810 mi/h. By how much should the tug's velocity be increased as it reaches *B* to insert it into the larger circular orbit?

12.44 A 3-oz ball slides on a smooth horizontal table at the end of a string which passes through a small hole in the table at *O*. When the length of string above the table is $r_1 = 15$ in., the speed of the ball is $v_1 = 8$ ft/s. Knowing that the breaking strength of the string is 3.00 lb, determine (*a*) the smallest distance r_2 which can be achieved by slowly drawing the string through the hole, (*b*) the corresponding speed v_2.

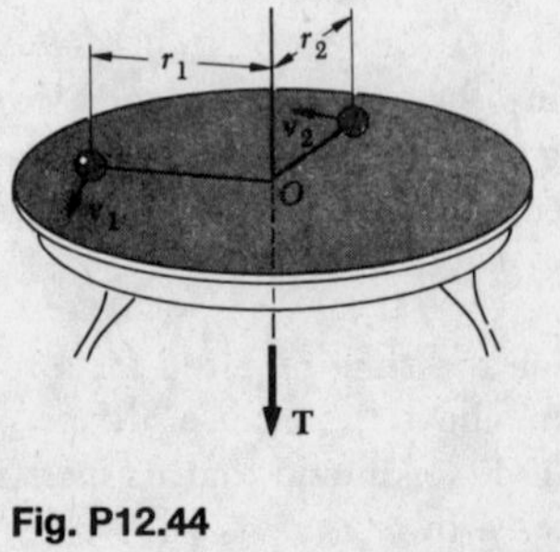

Fig. P12.44

SECTIONS 12.11 to 12.13

12.45 Plans for an unmanned landing mission on the planet Mars call for the earth-return vehicle to first describe a circular orbit about the planet. As it passes through point A, the vehicle will be inserted into an elliptic transfer orbit by firing its engine and increasing its speed by Δv_A. As it passes through point B, the vehicle will be inserted into a second transfer orbit located in a slightly different plane, by changing the direction of its velocity and by reducing its speed by Δv_B. Finally, as the vehicle passes through point C, its speed will be increased by Δv_C to insert it into its return trajectory. Knowing that the radius of the planet Mars is $R = 3400$ km, that its mass is 0.108 times the mass of the earth, and that the altitudes of points A and B are, respectively, $d_A = 2500$ km and $d_B = 90\,000$ km, determine the increase in speed Δv_A required at point A to insert the vehicle into its first transfer orbit.

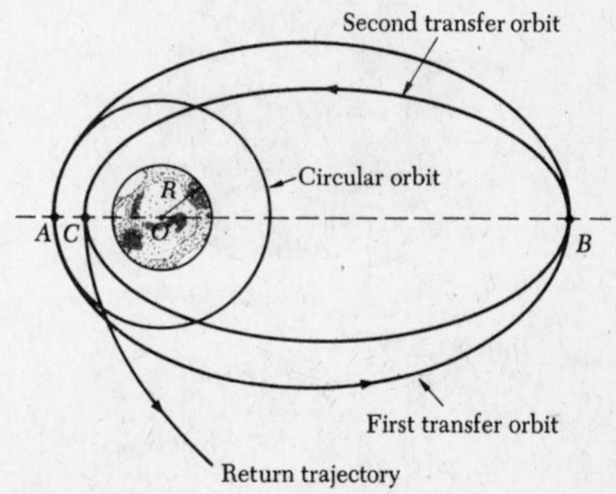

Fig. P12.45

12.46 For the vehicle of Prob. 12.45, it is known that the altitudes of points A, B, and C are, respectively, $d_A = 2500$ km, $d_B = 90\,000$ km, and $d_C = 1000$ km. Determine the change in speed Δv_B required at point B to insert the vehicle into its second transfer orbit.

12.47 A spacecraft is describing a circular orbit of 5000-mi radius around the earth when its engine is suddenly fired, increasing the speed of the spacecraft by 3000 mi/h. Determine the greatest distance from the center of the earth reached by the spacecraft.

12.48 A spacecraft is describing a circular orbit at an altitude of 240 mi above the surface of the earth when its engine is fired and its speed increased by 4000 ft/s. Determine the maximum altitude reached by the spacecraft.

12.49 For the vehicle of Prob. 12.45, it is known that the altitudes of points B and C are, respectively, $d_B = 90\,000$ km and $d_C = 1000$ km. Determine the minimum increase in speed Δv_C required at point C to insert the vehicle into an escape trajectory.

12.50 The maximum altitude a_1 of an unmanned earth satellite is observed to be 8300 km, while its minimum altitude a_2 is known to be 225 km. Determine the maximum and minimum values of its velocity.

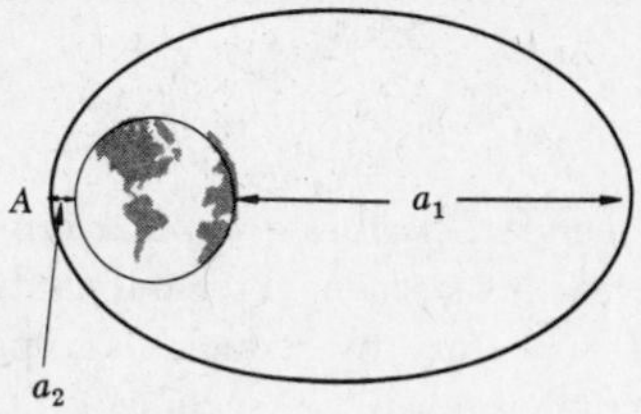

Fig. P12.50

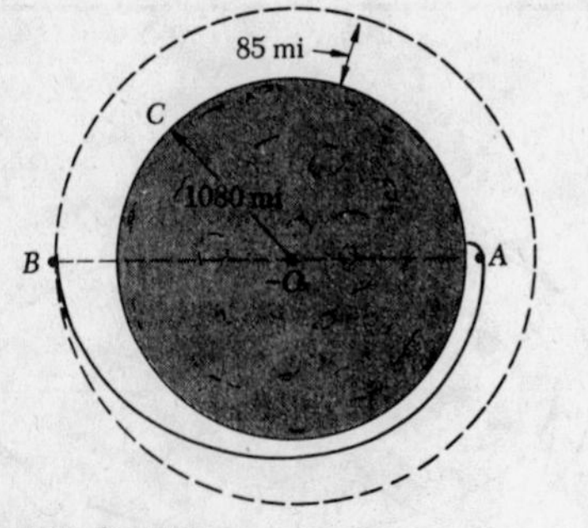

Fig. P12.51

12.51 After completing their moon-exploration mission, the two astronauts forming the crew of an Apollo lunar excursion module (LEM) prepare to rejoin the command module which is orbiting the moon at an altitude of 85 mi. They fire the LEM's engine, bring it along a curved path to a point A, 5 mi above the moon's surface, and shut off the engine. Knowing that the LEM is moving at that time in a direction parallel to the moon's surface and that it will coast along an elliptic path to a rendezvous at B with the command module, determine (*a*) the speed of the LEM at engine shutoff, (*b*) the relative velocity with which the command module will approach the LEM at B. (The radius of the moon is 1080 miles and its mass is 0.01230 times the mass of the earth.)

12.52 Solve Prob. 12.51, assuming that the Apollo command module is orbiting the moon at an altitude of 55 mi.

12.53 Referring to Prob. 12.45, determine the time required for the vehicle to describe its first transfer orbit from A to B.

12.54 Determine the periodic time of the satellite of Prob. 12.50.

12.55 Determine the time required for the LEM of Prob. 12.51 to travel from A to B.

12.56 Determine the approximate time required for an object to fall to the surface of the earth after being released with no velocity from a distance equal to the radius of the orbit of the moon, namely, 239,000 mi. (*Hint.* Assume that the object is given a very small initial velocity in a transverse direction, say, $v_\theta = 1$ ft/s, and determine the periodic time τ of the object on the resulting orbit. An examination of the orbit will show that the time of fall must be approximately equal to $\frac{1}{2}\tau$.)

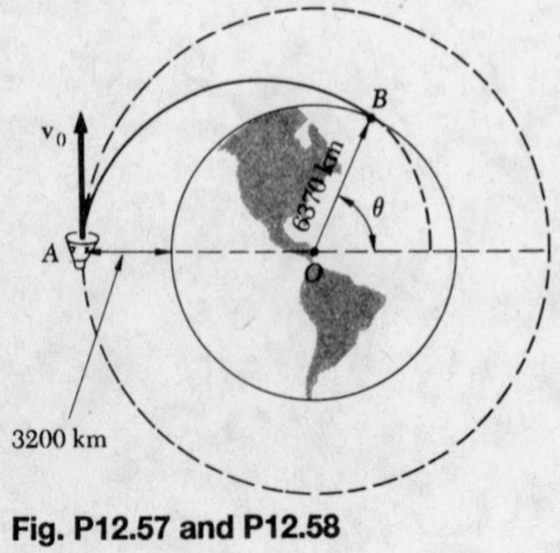

Fig. P12.57 and P12.58

12.57 A spacecraft describes a circular orbit at an altitude of 3200 km above the earth's surface. Preparatory to reentry it reduces its speed to a value $v_0 = 5400$ m/s, thus placing itself on an elliptic trajectory. Determine the value of θ defining the point B where splashdown will occur. (*Hint.* Point A is the apogee of the elliptic trajectory.)

12.58 A spacecraft describes a circular orbit at an altitude of 3200 km above the earth's surface. Preparatory to reentry it places itself on an elliptic trajectory by reducing its speed to a value v_0. Determine v_0 so that splashdown will occur at a point B corresponding to $\theta = 120°$. (See hint of Prob. 12.57.)

12.59 Upon the LEM's return to the command module, the Apollo spacecraft of Prob. 12.51 is turned around so that the LEM faces to the rear. After completing a full orbit, i.e., as the craft passes again through B, the LEM is cast adrift and crashes on the moon's surface at point C. Determine the velocity of the LEM relative to the command module as it is cast adrift, knowing that the angle BOC is 90°. (*Hint.* Point B is the apogee of the elliptic crash trajectory.)

12.60 Upon the LEM's return to the command module, the Apollo spacecraft of Prob. 12.51 is turned around so that the LEM faces to the rear. After completing a full orbit, i.e., as the craft passes again through B, the LEM is cast adrift with a velocity of 600 ft/s relative to the command module. Determine the point C where the LEM will crash on the moon's surface. (See hint of Prob. 12.59.)

CHAPTER 13
KINETICS OF PARTICLES: ENERGY AND MOMENTUM METHODS

SECTIONS 13.1 to 13.5

13.1 The conveyor belt shown moves at a constant speed v_0 and discharges packages on to the chute AB. The coefficient of friction between the packages and the chute is 0.50. Knowing that the packages must reach point B with a speed of 12 ft/s, determine the required speed v_0 of the conveyor belt.

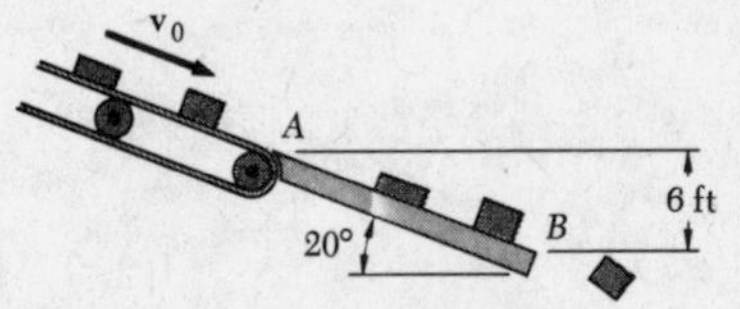

Fig. P13.1

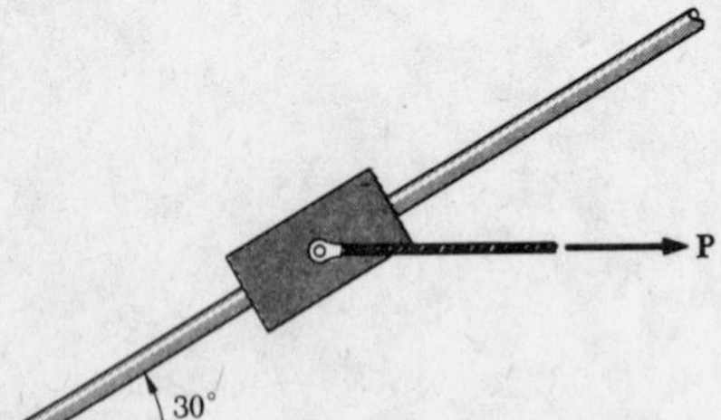

Fig. P13.2

13.2 The 2-kg collar was moving down the rod with a velocity of 3 m/s when a force **P** was applied to the horizontal cable. Assuming negligible friction between the collar and the rod, determine the magnitude of the force **P** if the collar stopped after moving 1.2 m more down the rod.

13.3 Solve Prob. 13.2, assuming a coefficient of friction of 0.20 between the collar and the rod.

13.4 A railroad car weighing 40 tons rolls 500 ft down a 2 per cent incline and then rolls 333 ft up a 2 per cent incline before coming to a stop. Determine the average rolling resistance of the car.

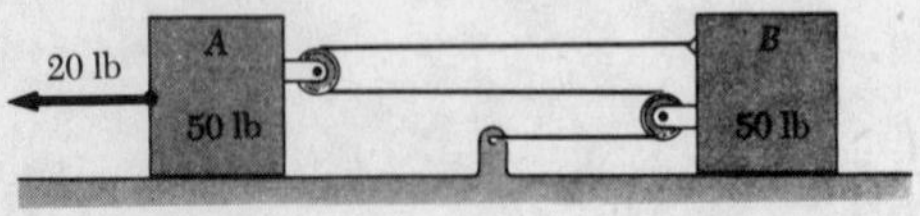

Fig. P13.5

13.5 The system shown is at rest when the 20-lb force is applied to block A. Neglecting the effect of friction, determine the velocity of block A after it has moved 9 ft.

13.6 Solve Prob. 13.5, assuming that the coefficient of friction between the blocks and the horizontal plane is 0.20.

13.7 Two blocks are joined by an inextensible cable as shown. If the system is released from rest, determine the velocity of block A after it has moved 2 m. Assume that μ equals 0.25 between block A and the plane and neglect the mass and friction of the pulleys.

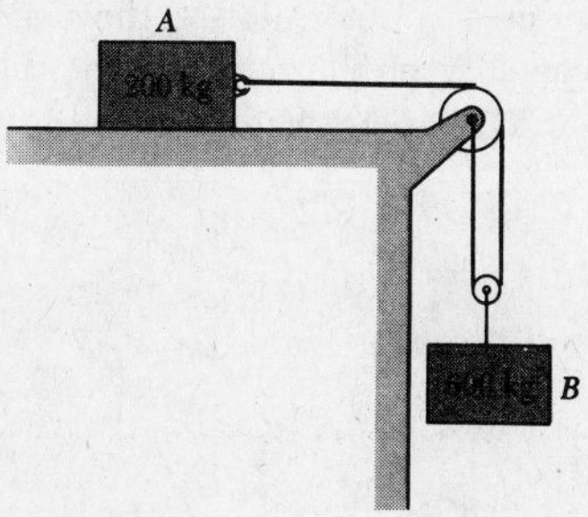

Fig. P13.7

13.8 Knowing that the system shown is initially at rest and neglecting the effect of friction, determine the force **P** required if the velocity of collar B is to be 3 m/s after it has moved 1.2 m to the right.

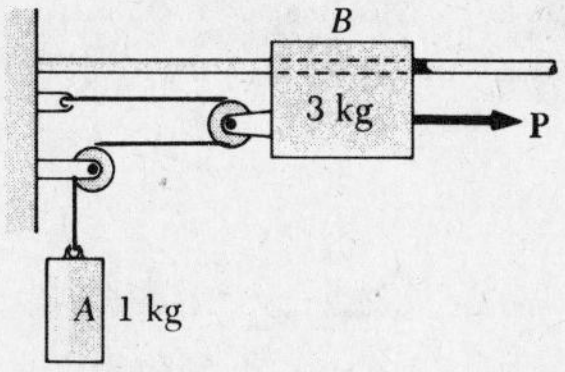

Fig. P13.8

13.9 Three 20-kg packages rest on a belt which passes over a pulley and is attached to a 40-kg block. Knowing that the coefficient of friction between the belt and the horizontal surface and also between the belt and the packages is 0.50, determine the speed of package B as it falls off the belt at E.

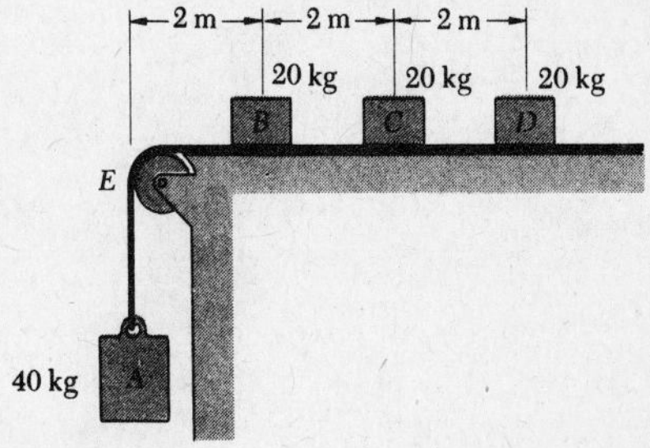

Fig. P13.9

13.10 In Prob. 13.9, determine the speed of package C as it falls off the belt at E.

13.11 Two cylinders are suspended from an inextensible cable as shown. If the system is released from rest, determine (*a*) the maximum velocity attained by the 10-lb cylinder, (*b*) the maximum height above the floor to which the 10-lb cylinder will rise.

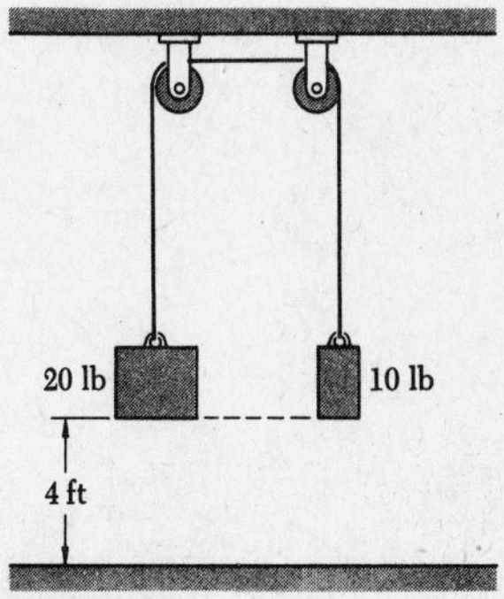

Fig. P13.11

13.12 Solve Prob. 13.11, assuming that the 20-lb cylinder is replaced by a 50-lb cylinder.

13.13 An industrial hoist can lift its maximum allowable load of 60,000 lb at the rate of 4 ft/min. Knowing that the hoist is run by a 15-hp motor, determine the overall efficiency of the hoist.

13.14 The escalator shown is designed to transport 9000 persons per hour at a constant speed of 90 ft/min. Assuming an average weight of 150 lb per person, determine (*a*) the average power required, (*b*) the required capacity of the motor if the mechanical efficiency is 75 percent and if a 250-percent overload is to be allowed.

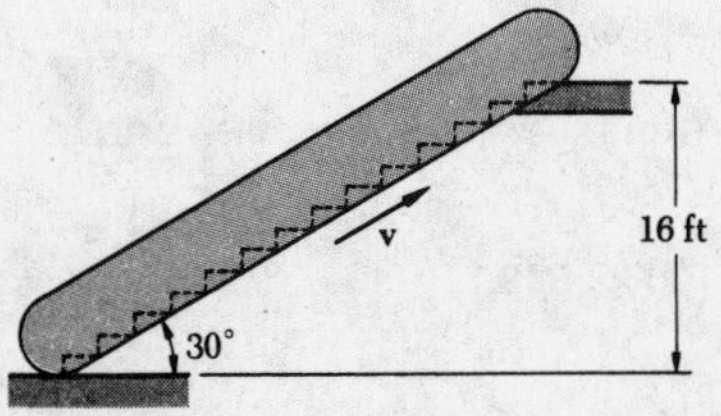

Fig. P13.14

13.15 The dumbwaiter *D* and its countermass *C* each have a mass of 350 kg. Determine the power required when the dumbwaiter (*a*) is moving upward at a constant speed of 4 m/s, (*b*) has an instantaneous velocity of 4 m/s upward and an upward acceleration of 0.9 m/s^2.

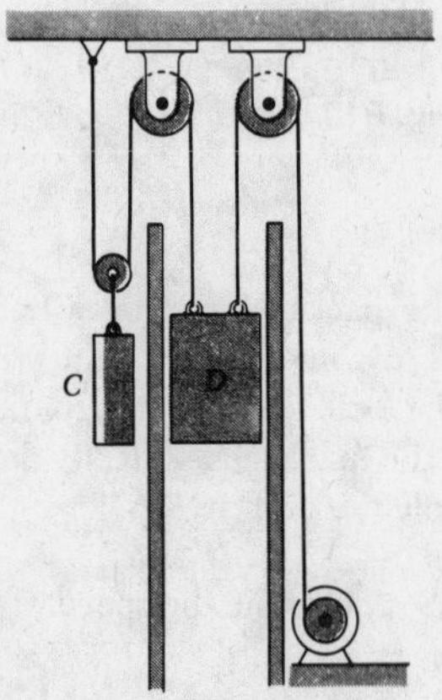

Fig. P13.15 and P13.16

13.16 The dumbwaiter *D* and its countermass *C* each have a mass of 350 kg. Knowing that the motor is delivering to the system 7 kW of power at the instant the speed of the dumbwaiter is 4 m/s upward, determine the acceleration of the dumbwaiter.

13.17 The spring AB is of constant 6 lb/in. and is attached to the 4-lb collar A which moves freely along the horizontal rod. The unstretched length of the spring is 10 in. If the collar is released from rest in the position shown, determine the maximum velocity attained by the collar.

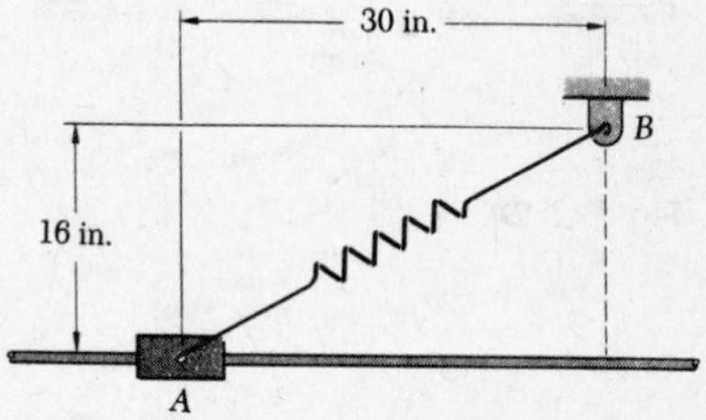

Fig. P13.17

13.18 In Prob. 13.17, determine the weight of the collar A for which the maximum velocity is 30 ft/s.

13.19 A collar of mass 1.5 kg is attached to a spring and slides without friction along a circular rod which lies in a *horizontal* plane. The spring is undeformed when the collar is at C and the constant of the spring is 400 N/m. If the collar is released from rest at B, determine the velocity of the collar as it passes through point C.

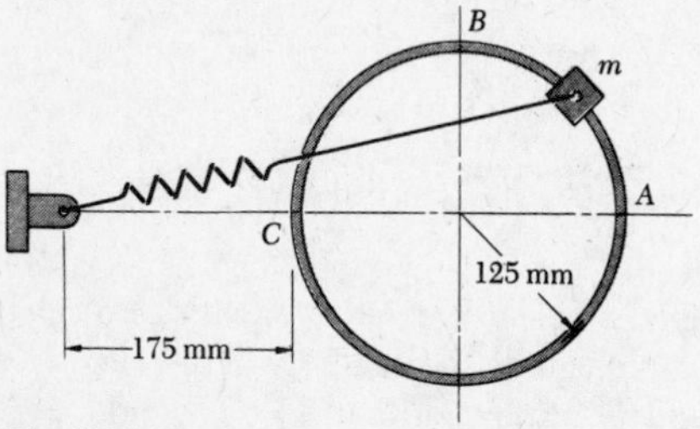

Fig. P13.19

13.20 The collar of Prob. 13.19 has a continuous, although non-uniform, motion along the rod. If the speed of the collar at A is to be half of its speed at C, determine (*a*) the required speed at C, (*b*) the corresponding speed at B.

13.21 The 50-kg block is released from rest when $\phi = 0$. If the speed of the block when $\phi = 90°$ is to be 2.5 m/s, determine the required value of the initial tension in the spring.

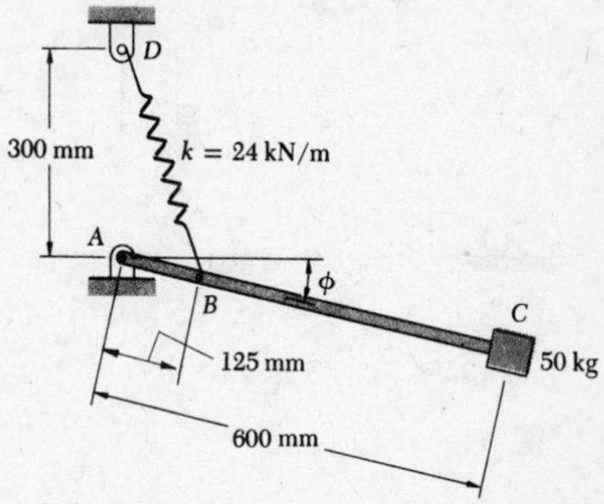

Fig. P13.21

13.22 The collar of Prob. 13.19 is released from rest at point A. Determine the horizontal component of the force exerted by the rod on the collar as the collar passes through point B. Show that the force component is independent of the mass of the collar.

13.23 A 1.5-lb collar may slide without friction along the semicircular rod BCD. The spring is of constant 2 lb/in. and its undeformed length is 12 in. The collar is released from rest at B. As the collar passes through point C, determine (*a*) the speed of the collar, (*b*) the force exerted by the rod on the collar.

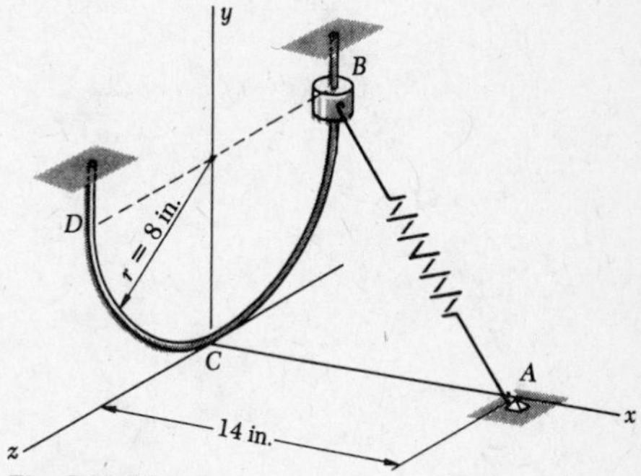

Fig. P13.23

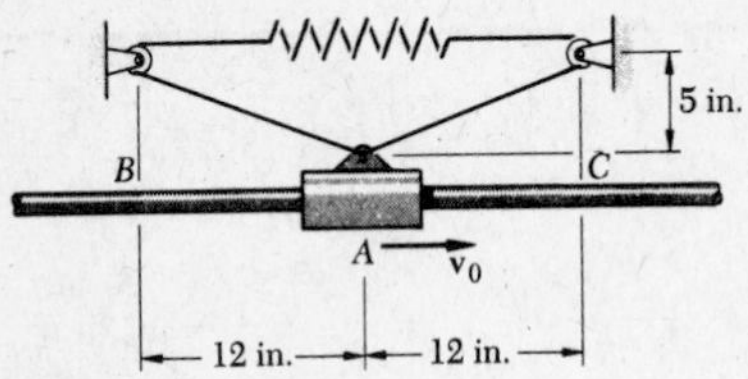

Fig. P13.24

13.24 The 2-lb collar slides without friction along the horizontal rod. Knowing that the constant of the spring is 3 lb/in. and that $v_0 = 12$ ft/s, determine the required spring tension in the position shown if the speed of the collar is to be 8 ft/s at point C.

13.25 A $\frac{1}{4}$-lb collar may slide without friction on a rod in a vertical plane. The collar is released from rest when the spring is compressed 1.5 in. As the collar passes through point B, determine (*a*) the speed of the collar, (*b*) the force exerted by the rod on the collar.

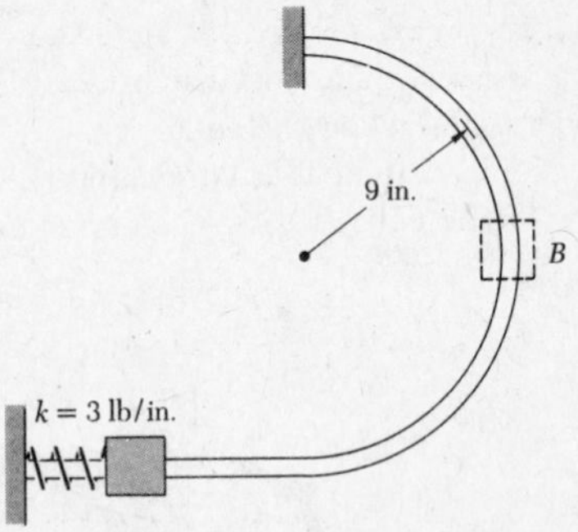

Fig. P13.25

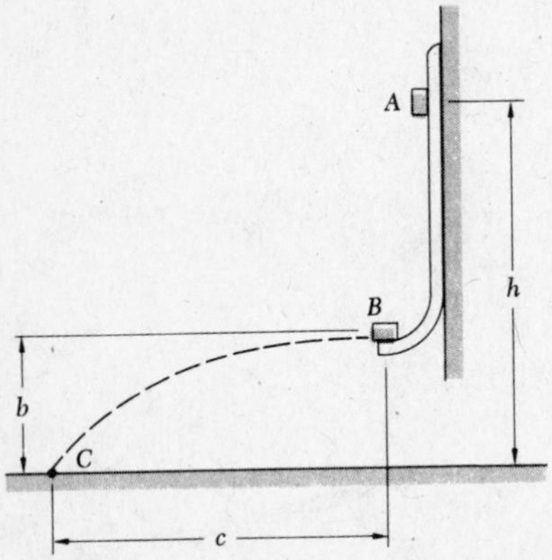

Fig. P13.26

13.26 A small block is released at A with zero velocity and moves along the frictionless guide to point B where it leaves the guide with a horizontal velocity. Knowing that $h = 8$ ft and $b = 3$ ft, determine (*a*) the speed of the block as it strikes the ground at C, (*b*) the corresponding distance c.

13.27 Assuming a given height h in Prob. 13.26, (*a*) show that the speed at C is independent of the height b, (*b*) determine the height b for which the distance c is maximum and the corresponding value of c.

13.28 A ball of mass m attached to an inextensible cord rotates in a vertical circle of radius r. Show that the difference between the maximum value T_{max} of the tension in the cord and its minimum value T_{min} is independent of the speed v_0 of the ball as measured at the bottom of the circle, and determine $T_{max} - T_{min}$.

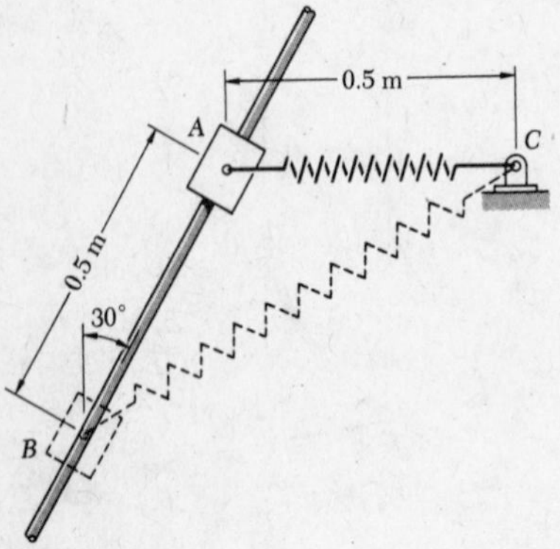

Fig. P13.29

13.29 A 5-kg collar slides without friction along a rod which forms an angle of 30° with the vertical. The spring is unstretched when the collar is at A. If the collar is released from rest at A, determine the value of the spring constant k for which the collar has zero velocity at B.

13.30 In Prob. 13.29, determine the value of the spring constant k for which the velocity of the collar at B is 1.5 m/s.

13.31 Prove that a force $\mathbf{F}(x,y,z)$ is conservative if, and only if, the following relations are satisfied:

$$\frac{\partial F_x}{\partial y} = \frac{\partial F_y}{\partial x} \qquad \frac{\partial F_y}{\partial z} = \frac{\partial F_z}{\partial y} \qquad \frac{\partial F_z}{\partial x} = \frac{\partial F_x}{\partial z}$$

13.32 The force $\mathbf{F} = (x\mathbf{i} + y\mathbf{j})/(x^2 + y^2)$ acts on the particle $P(x,y)$ which moves in the xy plane. (*a*) Using the first of the relations derived in Prob. 13.31, prove that $\mathbf{F}$ is a conservative force. (*b*) Determine the potential function $V(x,y)$ associated with $\mathbf{F}$.

13.33 The force $\mathbf{F} = (y + 2)\mathbf{i} + (2x - 2)\mathbf{j}$ acts on the particle $P(x,y)$ which moves in the xy plane. Determine the work of $\mathbf{F}$ knowing that P is initially at point A and moves along (*a*) the path ABC, (*b*) the path AC, (*c*) the path ADC.

13.34 The force $\mathbf{F} = x^2y\mathbf{i} + xy^2\mathbf{j}$ acts on the particle $P(x,y)$ which moves in the xy plane. Prove that $\mathbf{F}$ is a nonconservative force and determine the work of $\mathbf{F}$ as it moves from A to C along each of the paths ABC, ADC, and AC.

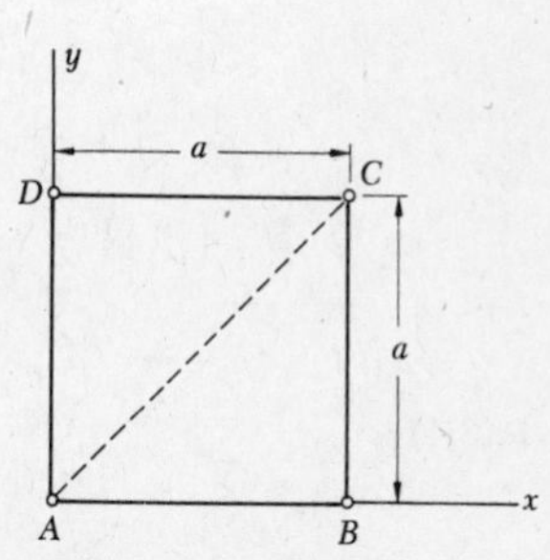

Fig. P13.33 and P13.34

13.35 How much energy per kilogram should be imparted to a satellite in order to place it in a circular orbit at an altitude of (*a*) 500 km, (*b*) 5000 km?

13.36 How much energy should be imparted to a 5-ton satellite in order to place it in a circular orbit at an altitude of (*a*) 300 miles, (*b*) 3,000 miles?

13.37 Determine the energy which must be imparted to a missile of mass m (*a*) to shoot it vertically to a height equal to the radius R of the earth, (*b*) to place it in a circular orbit of radius $2R$.

13.38 A 500-g block P rests on a frictionless horizontal table at a distance of 400 mm from a fixed pin O. The block is attached to pin O by an elastic cord of constant k = 100 N/m and of undeformed length 900 mm. If the block is set in motion to the right as shown, determine (*a*) the speed v_1 for which the distance from O to the block P will reach a maximum value of 1.2 m, (*b*) the speed v_2 when OP = 1.2 m, (*c*) the radius of curvature of the path of the block when OP = 1.2 m.

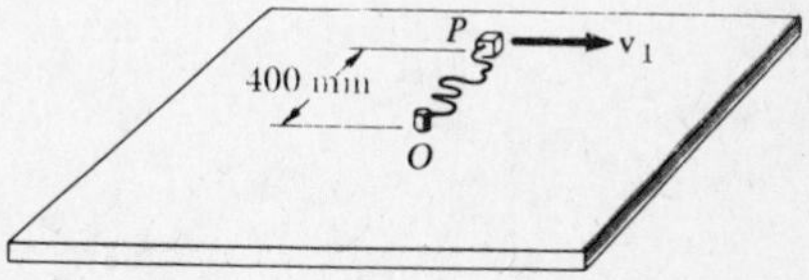

Fig. P13.38

13.39 A ball of mass m slides on a smooth horizontal table at the end of a string which passes through a small hole in the table at O. When the length of string above the table is r_1, the speed of the ball is v_1. If the string is drawn in until the length of string above the table is $r_2 = \frac{1}{2}r_1$, determine (*a*) the speed v_2, (*b*) the per cent change in the magnitude of the linear momentum of the ball, (*c*) the per cent change in the angular momentum of the ball with respect to O, (*d*) the per cent change in the kinetic energy of the ball.

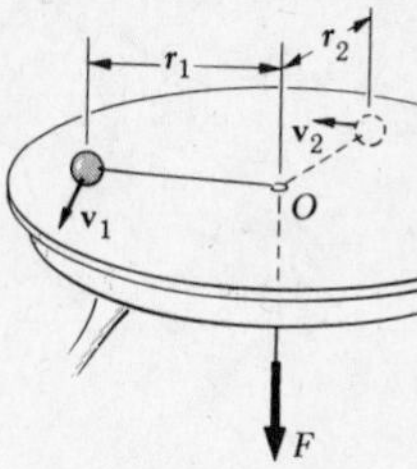

Fig. P13.39

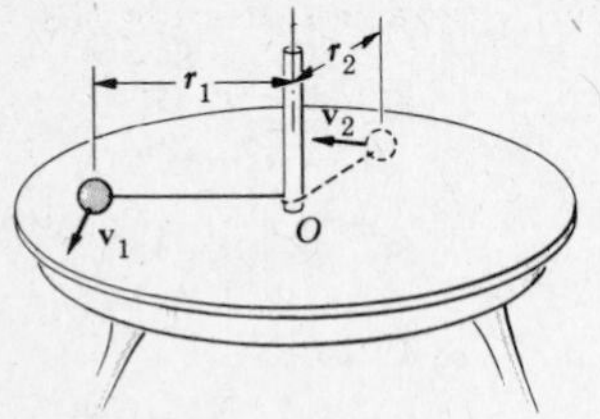

Fig. P13.40

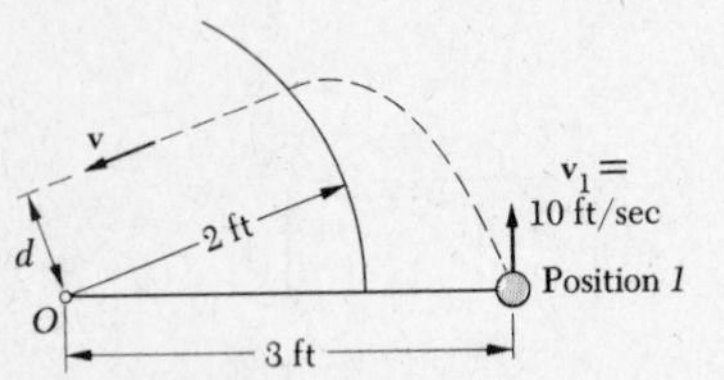

Fig. P13.41

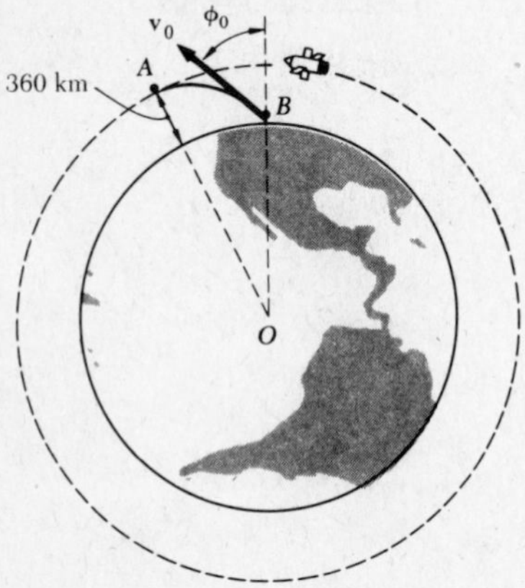

Fig. P13.46

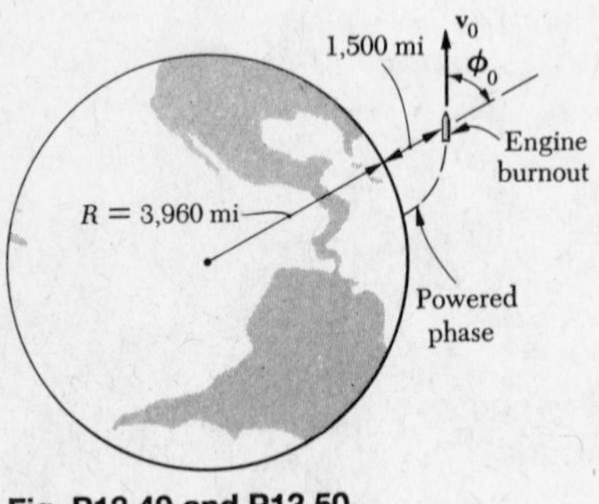

Fig. P13.49 and P13.50

13.40 Solve Prob. 13.39 assuming that the string wraps around a rod at O of very small diameter instead of being drawn through a hole at O. (*Hint.* No work is done by the forces acting on the system.)

13.41 A $\frac{1}{2}$-lb ball is attached to a fixed point O by means of an elastic cord of constant $k = 10$ lb/ft and of undeformed length 2 ft. The ball slides on a smooth horizontal surface. When the ball is in position *1*, the length of the cord is 3 ft and the velocity of the ball is 10 ft/sec, directed as shown. Determine (*a*) the speed of the ball after the cord has become slack, (*b*) the distance d when the ball is closest to point O.

13.42 In Prob. 13.41, determine the required magnitude of $\mathbf{v}_1$ if the ball is to pass at a distance $d = 4$ in. from point O. The direction of $\mathbf{v}_1$ is not changed.

13.42 through 13.45 Using the principles of conservation of energy and conservation of angular momentum, solve the following problems:

13.43 Prob. 12.45.
13.44 Prob. 12.51.
13.45 Prob. 12.50.

13.46 A space shuttle is to rendezvous with an orbiting laboratory which circles the earth at the constant altitude of 360 km. The shuttle has reached an altitude of 60 km and a velocity $\mathbf{v}_0$ of magnitude 3.5 km/s when its engine is shut off. What is the angle ϕ_0 that $\mathbf{v}_0$ should form with the vertical OB if the shuttle's trajectory is to be tangent at A to the orbit of the laboratory?

13.47 Determine the magnitude and direction (angle ϕ formed with the vertical OB) of the velocity $\mathbf{v}_B$ of the spacecraft of Prob. 12.57 just before splashdown at B. Neglect the effect of the atmosphere.

13.48 To what value v_0 should the speed of the spacecraft of Prob. 12.58 be reduced preparatory to reentry if its velocity $\mathbf{v}_B$ just before splashdown at B is to form an angle $\phi = 30°$ with the vertical OB? Neglect the effect of the atmosphere.

13.49 At engine burnout a satellite has reached an altitude of 1,500 miles and has a velocity $\mathbf{v}_0$ of 26,700 ft/sec forming an angle $\phi_0 = 76°$ with the vertical. Determine the maximum and minimum heights reached by the satellite.

13.50 At engine burnout a satellite has reached an altitude of 1,500 miles and has a velocity $\mathbf{v}_0$ of magnitude 26,700 ft/sec. For what range of values of the angle ϕ_0, formed by $\mathbf{v}_0$ and the vertical, will the satellite go into a permanent orbit? (Assume that if the satellite gets closer than 200 miles from the earth's surface, it will soon burn up.)

SECTIONS 13.10 and 13.11

13.51 A tugboat exerts a constant force of 25 tons on a 200,000-ton oil tanker. Neglecting the frictional resistance of the water, determine the time required to increase the speed of the tanker (*a*) from 1 mi/h to 2 mi/h, (*b*) from 2 mi/h to 3 mi/h.

13.52 A 60,000-ton ocean liner has an initial velocity of 2 mi/h. Neglecting the frictional resistance of the water, determine the time required to bring the liner to rest by using a single tugboat which exerts a constant force of 50,000 lb.

13.53 A 1250-kg automobile is moving at a speed of 75 km/h when the brakes are fully applied, causing all four wheels to skid. Determine the time required to stop the automobile (*a*) on concrete ($\mu = 0.80$), (*b*) on ice ($\mu = 0.10$).

13.54 A 1.5-kg particle is acted upon by a force **F** of magnitude $F = 70t^2$ (N) which acts in the direction of the unit vector $\lambda = \frac{2}{7}\mathbf{i} + \frac{3}{7}\mathbf{j} + \frac{6}{7}\mathbf{k}$. Knowing that the velocity of the particle at $t = 0$ is $\mathbf{v} = (120 \text{ m/s})\mathbf{j} - (75 \text{ m/s})\mathbf{k}$, determine the velocity when $t = 3$ s.

13.55 and 13.56 The initial velocity of the 50-kg car is 5 m/s to the left. Determine the time *t* at which the car has (*a*) no velocity, (*b*) a velocity of 5 m/s to the right.

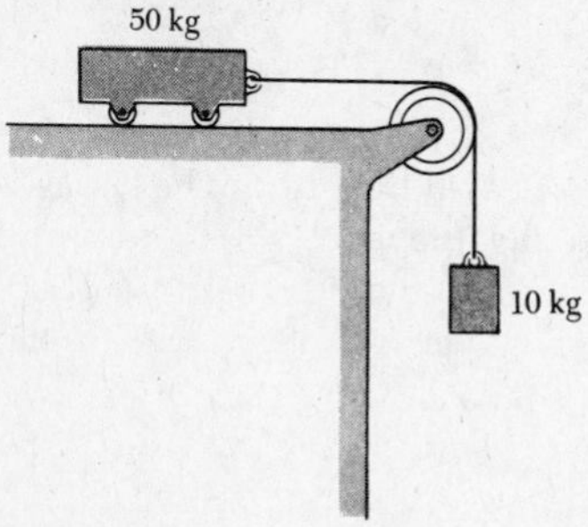

Fig. P13.55

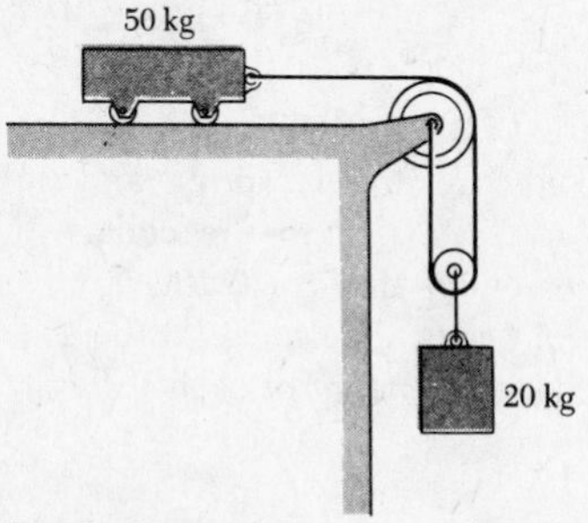

Fig. P13.56

13.57 A light train made of two cars travels at 60 mph. The first car weighs 15 tons, and the second car weighs 20 tons. When the brakes are applied, a constant braking force of 5000 lb is applied to each car. Determine (*a*) the time required for the train to stop after the brakes are applied, (*b*) the force in the coupling between the cars while the train is slowing down.

13.58 Solve Prob. 13.57, assuming that a constant braking force of 5000 lb is applied to car *B* but that the brakes on car *A* are not applied.

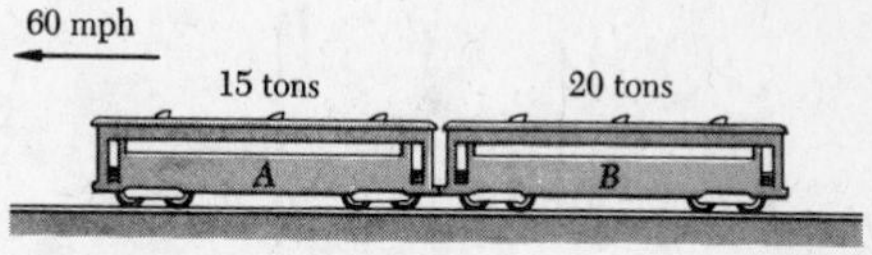

Fig. P13.57

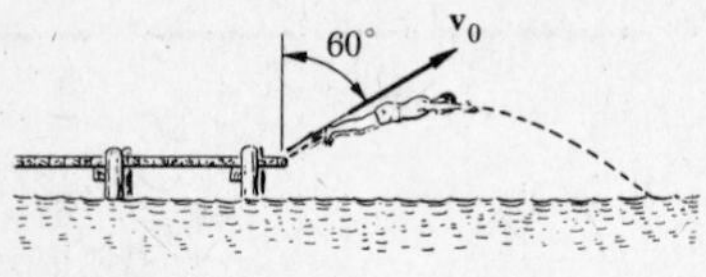

Fig. P13.59

13.59 A 70-kg man dives off the end of a pier with an initial velocity of 3 m/s in the direction shown. Determine the horizontal and vertical components of the average force exerted on the pier during the 0.8 s that the man takes to leave the pier.

13.60 Determine the initial recoil velocity of a 3.5-kg rifle which fires a 20-g bullet with a velocity of 500 m/s.

13.61 A 150-lb man dives horizontally off the end of a 300-lb boat, which is initially at rest. During the dive, the relative horizontal velocity of the man with respect to the boat is 12 ft/sec to the right. (*a*) Determine the resulting velocity of the boat. (*b*) If the man leaves the boat in 0.75 sec, determine the average impulsive force that he exerted on the boat.

13.62 A 50-g rifle bullet is fired horizontally with a velocity of 350 m/s into a 3.5-kg block of wood which can move freely in the horizontal direction. Determine (*a*) the final velocity of the block, (*b*) the ratio of the final kinetic energy of the block and bullet to the initial kinetic energy of the bullet.

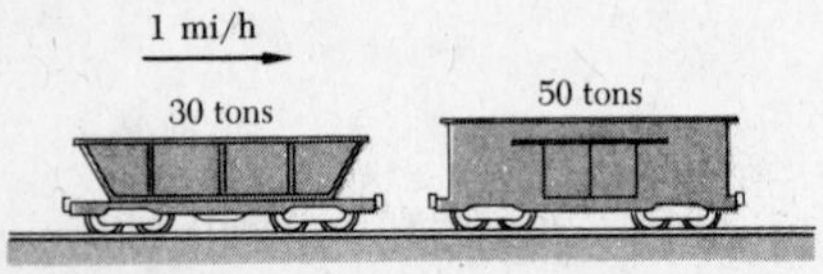

Fig. P13.63

13.63 A 30-ton railroad car is to be coupled to a second car which weighs 50 tons. If initially the speed of the 30-ton car is 1 mi/h and the 50-ton car is at rest, determine (*a*) the final speed of the coupled cars, (*b*) the average impulsive force acting on each car if the coupling is completed in 0.40 s.

13.64 Solve Prob. 13.63, assuming that, initially, the 30-ton car is at rest and the 50-ton car has a speed of 1 mi/h.

13.65 Collars *A* and *B* are moved toward each other, thus compressing the spring, and are then released from rest. The spring is not attached to the collars. Neglecting the effect of friction and knowing that collar *B* is observed to move to the right with a velocity of 6 m/s, determine (*a*) the corresponding velocity of collar *A*, (*b*) the kinetic energy of each collar.

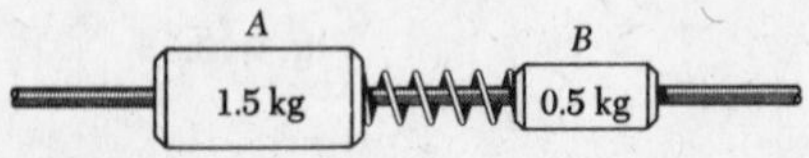

Fig. P13.65

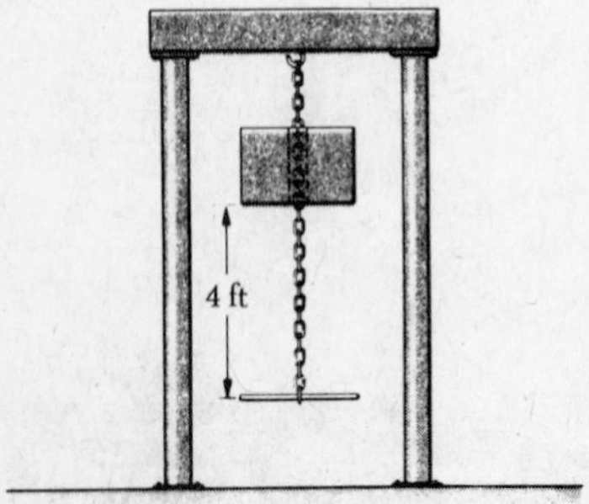

Fig. P13.66

13.66 In order to test the resistance of a chain to impact, the chain is suspended from a 200-lb dead weight supported by two columns. A rod attached to the last link of the chain is then hit by a 50-lb block dropped from a 4-ft height. Determine the initial impulse exerted on the chain, assuming that the impact is perfectly plastic and that the columns supporting the dead weight (*a*) are perfectly rigid, (*b*) are equivalent to two perfectly elastic springs. (*c*) Determine the energy absorbed by the chain in parts *a* and *b*.

SECTIONS 13.12 to 13.15

13.67 Knowing that $e = 0.75$, determine how much additional weight should be attached to collar B if its velocity after impact is to be zero.

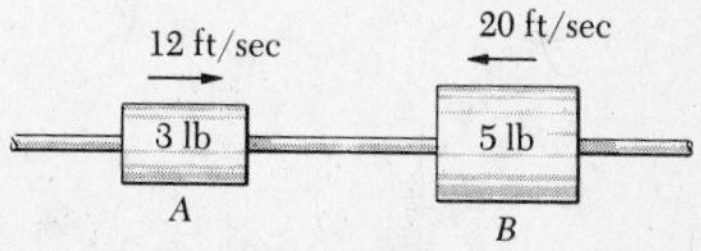

Fig. P13.67 and P13.68

13.68 The velocities of the two collars before impact are as shown. If after the impact the velocity of collar B is observed to be 2 ft/sec to the left, determine the coefficient of restitution between the two collars.

13.69 The coefficient of restitution between the two collars is known to be 0.75; determine (*a*) their velocities after impact, (*b*) the energy loss during the impact.

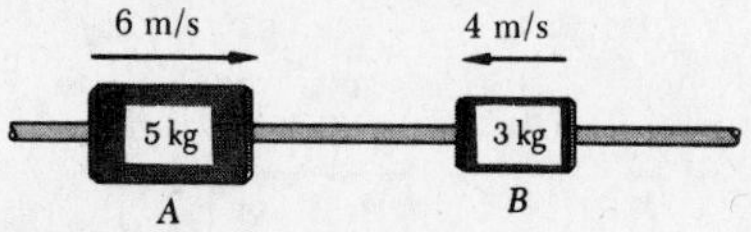

Fig. P13.69

13.70 Solve Prob. 13.69, assuming that the velocity of collar B is 2 m/s to the right.

13.71 Assuming perfectly elastic impact, determine the velocity imparted to a quarter-dollar coin which is at rest and is struck squarely by (*a*) a dime moving with a velocity $\mathbf{v}_0$, (*b*) a half-dollar moving with a velocity $\mathbf{v}_0$. (*Masses:* half-dollar, 12.50 g; quarter dollar, 6.25 g; dime, 2.50 g.)

13.72 In Sample Prob. 13.15, determine the required value of the coefficient of restitution if the angles α and β between the horizontal and the respective directions of the final velocities are to be equal.

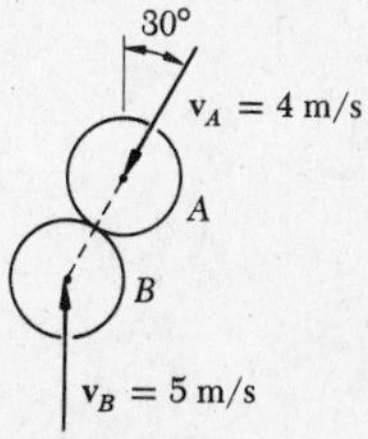

Fig. P13.73

13.73 As ball A is falling, a juggler tosses an identical ball B which strikes ball A. The line of impact forms an angle of 30° with the vertical. Assuming the balls frictionless and $e = 0.8$, determine the velocity of each ball immediately after impact.

13.74 A steel ball strikes a 90° corner at B, is deflected, and again strikes the corner at C. The coefficient of restitution is denoted by e. If the ball strikes B with a velocity of magnitude v, show (*a*) that the magnitude of its final velocity after striking C is ev and (*b*) that the initial and final paths AB and CD are parallel. Assume that v is large and that the short intermediate path BC is a straight line.

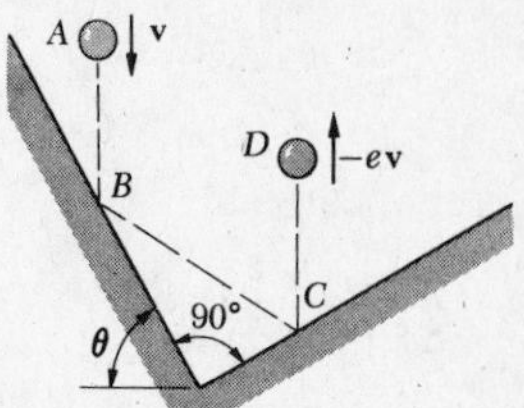

Fig. P13.74

13.75 In Prob. 13.74, derive an expression for θ, knowing that path BC is to be perpendicular to paths AB and CD.

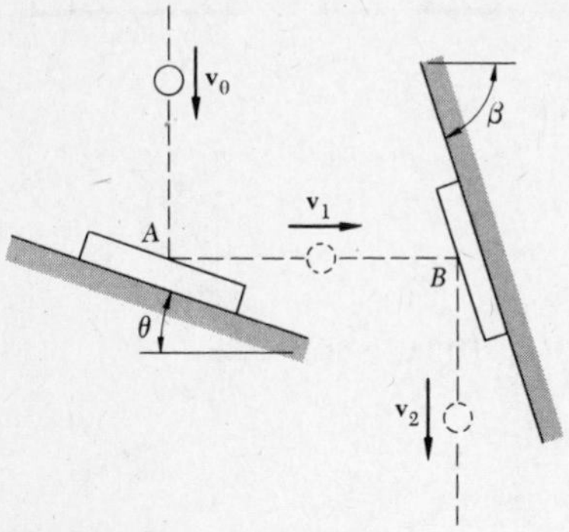

Fig. P13.76

13.76 A steel ball falling vertically strikes a rigid plate A and rebounds horizontally as shown. Denoting by e the coefficient of restitution, determine (*a*) the required angle θ, (*b*) the magnitude of the velocity v_1.

13.77 The ball of Prob. 13.76 strikes a second rigid plate B and rebounds vertically as shown. Determine (*a*) the required angle β, (*b*) the magnitude of the velocity v_2 with which the ball leaves plate B.

13.78 A ball is dropped from a height $h_0 = 900$ mm onto a frictionless floor. Knowing that for the first bounce $h_1 = 800$ mm and $d_1 = 400$ mm, determine (*a*) the coefficient of restitution, (*b*) the height and length of the second bounce.

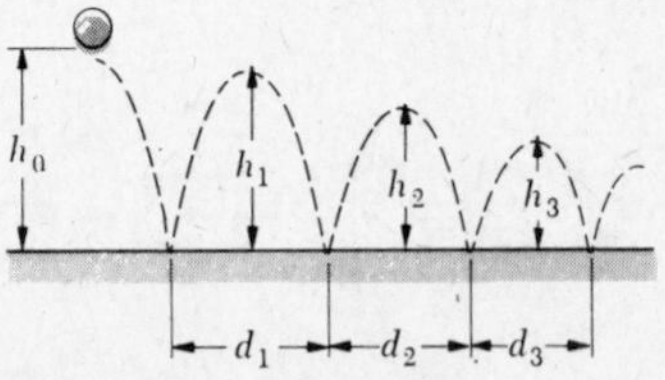

Fig. P13.78 and P13.79

13.79 A ball is dropped onto a frictionless floor and bounces as shown. The lengths of the first two bounces are measured and found to be $d_1 = 14.5$ in. and $d_2 = 12.8$ in. Determine (*a*) the coefficient of restitution, (*b*) the expected length d_3 of the third bounce.

13.80 The 4.5-kg sphere A strikes the 1.5-kg sphere B. Knowing that $e = 0.90$, determine the angle θ_A at which A must be released if the maximum angle θ_B reached by B is to be 90°.

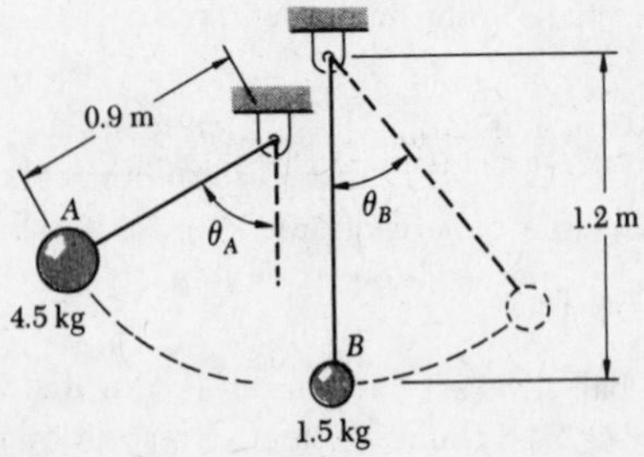

Fig. P13.80 and P13.81

13.81 The 4.5-kg sphere A is released from rest when $\theta_A = 60°$ and strikes the 1.5-kg sphere B. Knowing that $e = 0.90$, determine (*a*) the highest position to which sphere B will rise, (*b*) the maximum tension which will occur in the cord holding B.

13.82 Block A is released when $\theta_A = 90°$ and slides without friction until it strikes ball B. Knowing that $e = 0.90$, determine (*a*) the velocity of B immediately after impact, (*b*) the maximum tension in the cord holding B, (*c*) the maximum height to which ball B will rise.

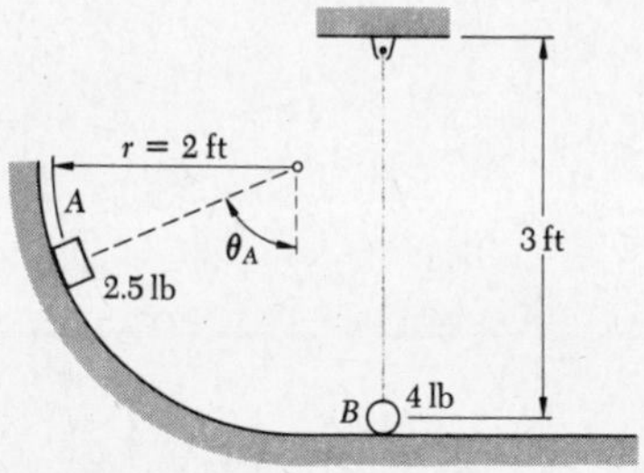

Fig. P13.82

13.83 What should be the value of the angle θ_A in Prob. 13.82 if the maximum angle between the cord holding ball B and the vertical is to be 45°?

13.84 The 4-lb sphere is released from rest when $\theta_A = 60°$. It is observed that the velocity of the sphere is zero after the impact and that the block moves 3 ft before coming to rest. Determine (*a*) the coefficient of restitution between the sphere and block, (*b*) the coefficient of friction between the block and the horizontal surface.

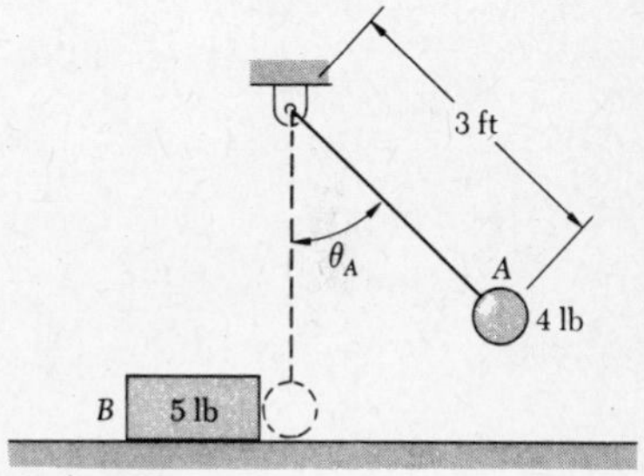

Fig. P13.84

13.85 Collar B has an initial velocity of 2 m/s. It strikes collar A causing a series of impacts involving the collars and the fixed support at C. Assuming $e = 1$ for all impacts and neglecting friction, determine (*a*) the number of impacts which will occur, (*b*) the final velocity of B, (*c*) the final position of A.

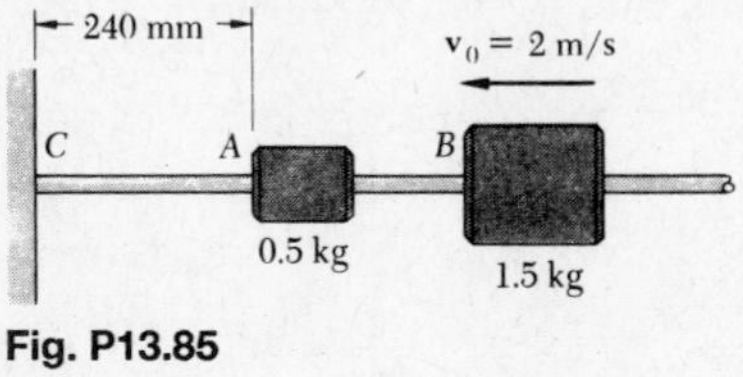

Fig. P13.85

13.86 Three steel rods of uniform cross section may slide without friction on a horizontal surface. Rods B and C are at rest when B is struck by rod A, which was moving with a velocity $\mathbf{v}_0$. (*a*) Denoting by e the coefficient of restitution of the rods, determine the length x of rod B for which we obtain the maximum velocity of rod C after it has been struck by B for the first time. (*b*) Using the value of x found in part *a*, determine the kinetic energy of C after impact as a fraction of the initial kinetic energy of A.

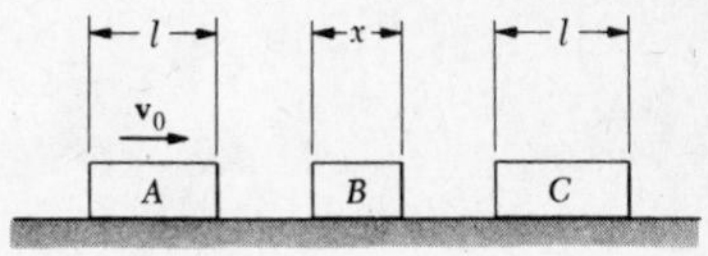

Fig. P13.86

CHAPTER 14
SYSTEMS OF PARTICLES

SECTIONS 14.1 to 14.6

14.1 Two men dive horizontally and to the right off the end of a 300-lb boat. The boat is initially at rest, and each man weighs 150 lb. If each man dives so that his relative horizontal velocity with respect to the boat is 12 ft/sec, determine (*a*) the velocity of the boat after the men dive simultaneously, (*b*) the velocity of the boat after one man dives and the velocity of the boat after the second man dives.

14.2 Solve Prob. 14.1, assuming that the first man dives to the right off the front end of the boat and the second man dives to the left off the back of the boat.

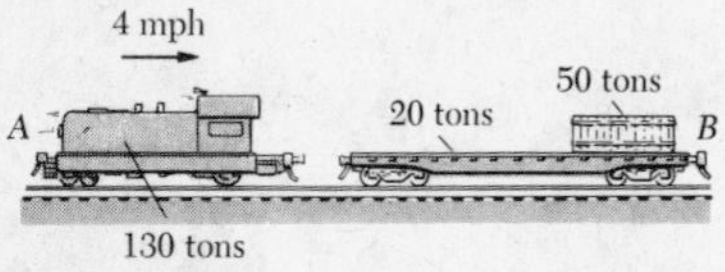

Fig. P14.3

14.3 A 130-ton engine coasting at 4 mph strikes, and is automatically coupled with, a 20-ton flat car which carries a 50-ton load. The load is *not* securely fastened to the car but may slide along the floor ($\mu = 0.20$). Knowing that the car was at rest with its brakes released and that the coupling takes place instantaneously, determine the velocity of the engine (*a*) immediately after the coupling, (*b*) after the load has slid to a stop relative to the car.

14.4 Solve Prob. 14.3, assuming that the coupling takes place in a period of 0.40 sec.

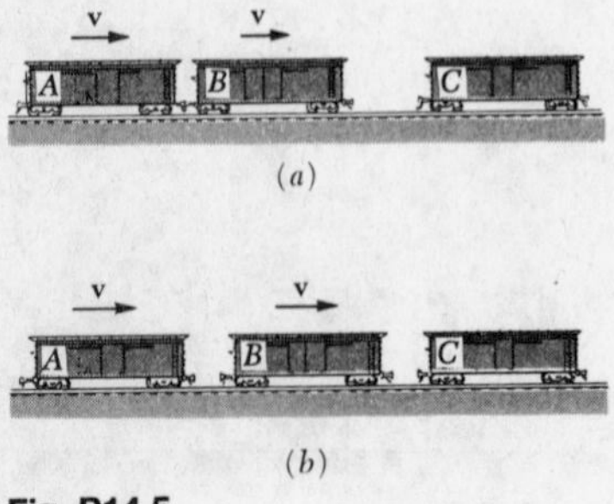

Fig. P14.5

14.5 Two railroad freight cars move with a velocity v through a switchyard. Car *B* hits a third car *C*, which was at rest with its brakes released, and it automatically couples with *C*. Knowing that all three cars have the same mass, determine their common velocity after they are all coupled together, as well as the percentage of their total initial kinetic energy which is absorbed by each coupling mechanism, assuming (*a*) that cars *A* and *B* were originally coupled, (*b*) that cars *A* and *B* were moving a few feet apart and that the coupling operation between *B* and *C* is completed before *A* hits *B* and becomes coupled with it.

14.6 Two identical balls B and C are at rest when ball B is struck by a ball A of the same mass, moving with a velocity of 4 m/s. This causes a series of collisions between the various balls. Knowing that $e = 0.40$, determine the velocity of each ball after *all* collisions have taken place.

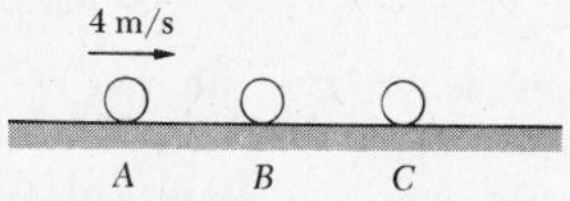

Fig. P14.6

14.7 A system consists of three particles A, B, and C. We know that $m_A = 1$ kg, $m_B = 1.5$ kg, and $m_C = 2$ kg and that the velocities of the particles expressed in metres per second are, respectively, $\mathbf{v}_A = -10\mathbf{j} + 5\mathbf{k}$, $\mathbf{v}_B = 8\mathbf{i} - 6\mathbf{j} + 4\mathbf{k}$, and $\mathbf{v}_C = v_x\mathbf{i} + v_y\mathbf{j} + 10\mathbf{k}$. Determine (*a*) the components v_x and v_y of the velocity of particle C for which the angular momentum $\mathbf{H}_O$ of the system about O is parallel to the z axis, (*b*) the corresponding value of $\mathbf{H}_O$.

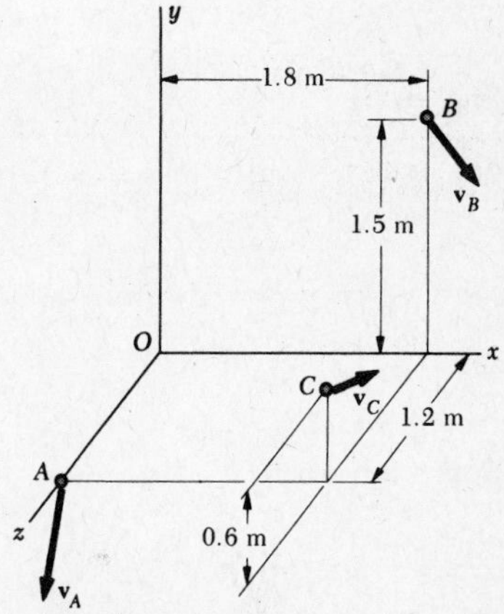

Fig. P14.7

14.8 For the system of particles of Prob. 14.7, determine (*a*) the components v_x and v_y of the velocity of particle C for which the angular momentum $\mathbf{H}_O$ of the system about O is parallel to the x axis, (*b*) the corresponding value of $\mathbf{H}_O$.

14.9 A 10-kg projectile is passing through the origin O with a velocity $\mathbf{v}_0 = (60\ \text{m/s})\mathbf{i}$ when it explodes into two fragments, A and B, of mass 4 kg and 6 kg, respectively. Knowing that, 2 s later, the position of the first fragment is $A(150\ \text{m},\ 12\ \text{m},\ -24\ \text{m})$, determine the position of fragment B at the same instant. Assume $g = 9.81\ \text{m/s}^2$ and neglect the resistance of the air.

14.10 A 300-lb space vehicle traveling with a velocity $\mathbf{v}_0 = 1{,}000\mathbf{k}$ (ft/sec) passes through the origin O at $t = 0$. Explosive charges then separate the vehicle into three parts A, B, and C, weighing, respectively, 50 lb, 100 lb, and 150 lb. Knowing that, at $t = 2$ sec, the positions of parts B and C are observed to be $\mathbf{r}_B = 500\mathbf{i} + 1{,}100\mathbf{j} + 2{,}700\mathbf{k}$ (ft) and $\mathbf{r}_C = -400\mathbf{i} - 800\mathbf{j} + 1{,}600\mathbf{k}$ (ft), determine the corresponding position of part A.

14.11 In Prob. 14.10 it is observed, at $t = 2$ sec, that the velocity of part C of the space vehicle is $\mathbf{v}_C = -200\mathbf{i} - 400\mathbf{j} + 800\mathbf{k}$ (ft/sec) and that the velocity of part B is $\mathbf{v}_B = (v_B)_x\mathbf{i} + 700\mathbf{j} + (v_B)_z\mathbf{k}$ (ft/sec). Determine the corresponding velocity of part A.

14.12 An archer hits a game bird flying in a horizontal straight line 30 ft above the ground with a 500-grain wooden arrow [1 grain = (1/7000) lb]. Knowing that the arrow strikes the bird from behind with a velocity of 350 ft/s at an angle of 30° with the vertical, and that the bird falls to the ground in 1.5 s and 48 ft beyond the point where it was hit, determine (*a*) the weight of the bird, (*b*) the speed at which it was flying when it was hit.

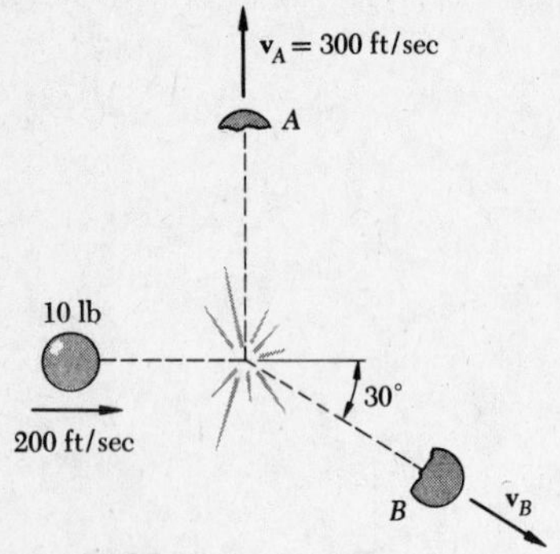

Fig. P14.13

14.13 A 10-lb sphere is moving with a velocity of 200 ft/sec when it explodes into two fragments. Immediately after the explosion the fragments are observed to travel in the directions shown and the speed of fragment *A* is observed to be 300 ft/sec. Determine (*a*) the weight of fragment *A*, (*b*) the speed of fragment *B*.

14.14 In a game of billiards, ball *A* is moving with the velocity $\mathbf{v}_0 = (2.5\text{ m/s})\mathbf{i}$ when it strikes balls *B* and *C* which are at rest side by side. After the collision, *A* is observed to move with the velocity $\mathbf{v}_A = (0.98\text{ m/s})\mathbf{i} - (1.14\text{ m/s})\mathbf{j}$, while *B* and *C* move in the directions shown. Determine the magnitudes of the velocities $\mathbf{v}_B$ and $\mathbf{v}_C$.

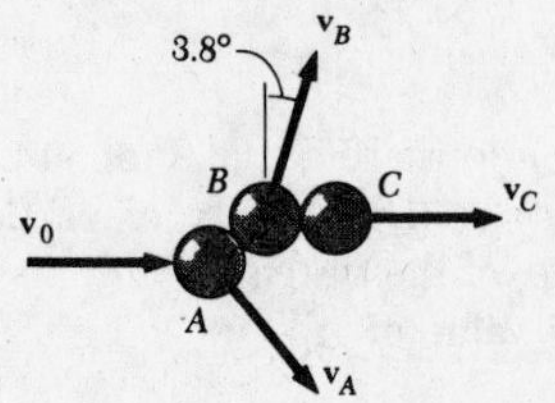

Fig. P14.14

14.15 A 5-kg object is falling vertically when, at point *D*, it explodes into three fragments *A*, *B*, and *C*, weighing, respectively, 1.5 kg, 2.5 kg, and 1 kg. Immediately after the explosion the velocity of each fragment is directed as shown and the speed of fragment *A* is observed to be 70 m/s. Determine the velocity of the 5-kg object immediately before the explosion.

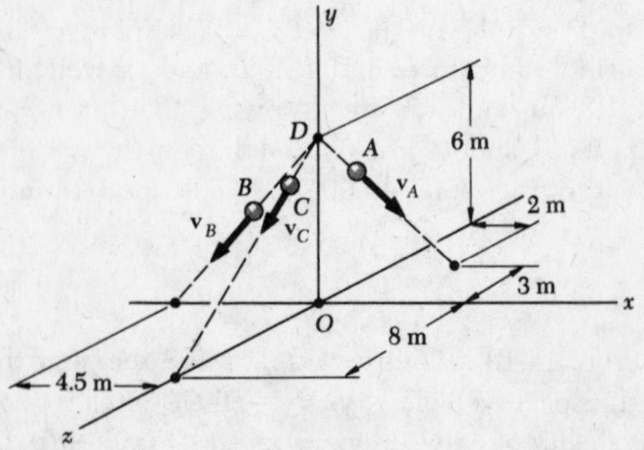

Fig. P14.15

SECTIONS 14.7 to 14.9

14.16 In Prob. 14.12, determine the amount of energy lost as the arrow hits the game bird.

14.17 In Prob. 14.14, determine the percentage of the initial kinetic energy lost due to the impacts among the three balls.

14.18 In Prob. 14.15, determine the work done by the internal forces during the explosion.

14.19 In Sample Prob. 14.2, determine the work done by the internal forces as the projectile explodes into the two fragments.

14.20 A 5-lb weight slides without friction on the xy plane. At $t = 0$ it passes through the origin with a velocity $\mathbf{v}_0 = (20 \text{ ft/s})\mathbf{i}$. Internal springs then separate the weight into the three parts shown. Knowing that, at $t = 3$ s, $\mathbf{r}_A = (42 \text{ ft})\mathbf{i} + (27 \text{ ft})\mathbf{j}$ and $\mathbf{r}_B = (60 \text{ ft})\mathbf{i} - (6 \text{ ft})\mathbf{j}$, that $\mathbf{v}_A = (14 \text{ ft/s})\mathbf{i} + (9 \text{ ft/s})\mathbf{j}$, and that $\mathbf{v}_B$ is parallel to the x axis, determine the corresponding position and velocity of part C.

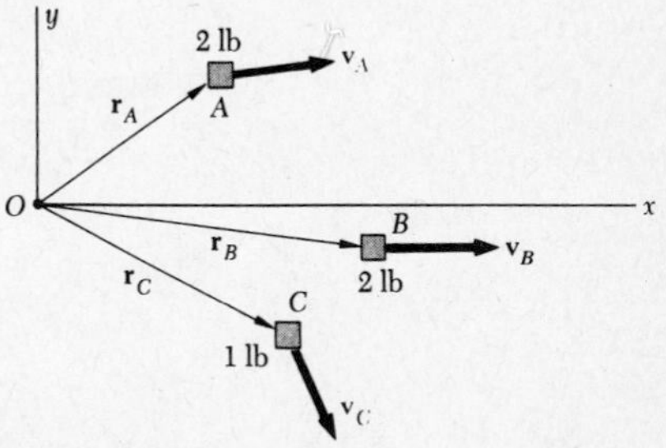

Fig. P14.20

14.21 In a game of billiards, ball A is moving with the velocity $\mathbf{v}_0 = v_0\mathbf{i}$ when it strikes balls B and C which are at rest side by side. After the collision, the three balls are observed to move in the directions shown. Assuming frictionless surfaces and perfectly elastic impacts (i.e., conservation of energy), determine the magnitudes of the velocities $\mathbf{v}_A$, $\mathbf{v}_B$, and $\mathbf{v}_C$ in terms of v_0 and θ.

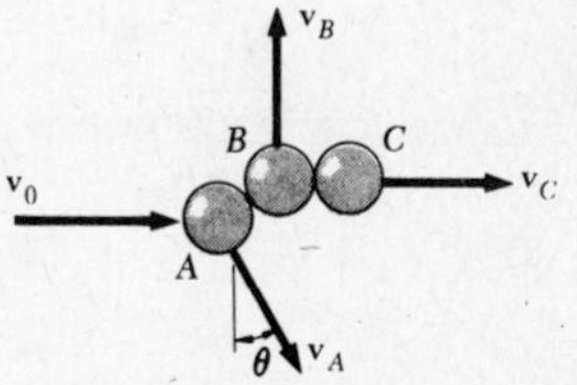

Fig. P14.21 and P14.22

14.22 In a game of billiards, ball A is moving with the velocity $\mathbf{v}_0 = (3 \text{ m/s})\mathbf{i}$ when it strikes balls B and C which are at rest side by side. After the collision, the three balls are observed to move in the directions shown, with $\theta = 30°$. Assuming frictionless surfaces and perfectly elastic impacts (i.e., conservation of energy), determine the magnitudes of the velocities $\mathbf{v}_A$, $\mathbf{v}_B$, and $\mathbf{v}_C$.

14.23 A 240-kg space vehicle traveling with the velocity $\mathbf{v}_0 = (500 \text{ m/s})\mathbf{k}$ passes through the origin O at $t = 0$. Explosive charges then separate the vehicle into three parts, A, B, and C, of mass 40 kg, 80 kg, and 120 kg, respectively. Knowing that at $t = 3$ s the positions of the three parts are, respectively, $A(150, 150, 1350)$, $B(375, 825, 2025)$, and $C(-300, -600, 1200)$, where the coordinates are expressed in meters, that the velocity of C is $\mathbf{v}_C = -(100 \text{ m/s})\mathbf{i} - (200 \text{ m/s})\mathbf{j} + (400 \text{ m/s})\mathbf{k}$, and that the y component of the velocity of B is $+350$ m/s, determine the velocity of part A.

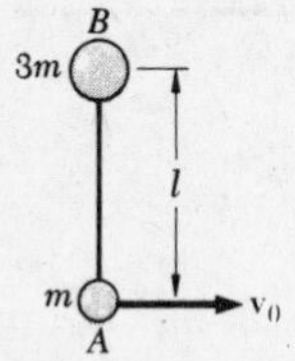

Fig. P14.24

14.24 Two small spheres A and B, respectively of mass m and $3m$, are connected by a rigid rod of length l and negligible mass. The two spheres are resting on a horizontal, frictionless surface when A is suddenly given the velocity $\mathbf{v}_0 = v_0\mathbf{i}$. Determine (*a*) the linear momentum of the system and its angular momentum about its mass center G, (*b*) the velocities of A and B after the rod AB has rotated through 90°, (*c*) the velocities of A and B after the rod AB has rotated through 180°.

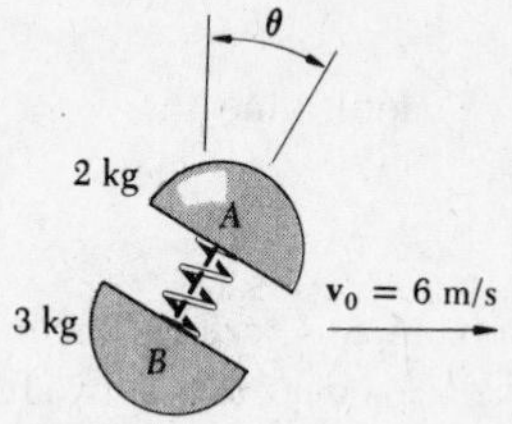

Fig. P14.25

14.25 When the cord connecting particles A and B is severed, the compressed spring causes the particles to fly apart (the spring is not connected to the particles). The potential energy of the compressed spring is known to be 60 J and the assembly has an initial velocity $\mathbf{v}_0$ as shown. If the cord is severed when $\theta = 30°$, determine the resulting velocity of each particle.

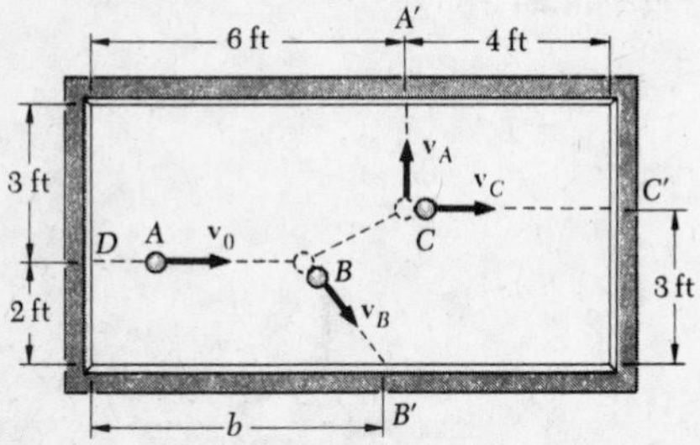

Fig. P14.26

14.26 In a game of billiards, ball A is given an initial velocity $\mathbf{v}_0$ along line DA parallel to the axis of the table. It hits ball B and then ball C, which are at rest. Knowing that A and C hit the sides of the table squarely at points A' and C', respectively, with velocities of magnitude $v_A = 4$ ft/s and $v_C = 6$ ft/s, and assuming frictionless surfaces and perfectly elastic impacts (i.e., conservation of energy), determine (*a*) the initial velocity $\mathbf{v}_0$ of ball A, (*b*) the velocity $\mathbf{v}_B$ of ball B, (*c*) the point B' where B hits the side of the table.

14.27 Solve Prob. 14.26 if $v_A = 6$ ft/s and $v_C = 4$ ft/s.

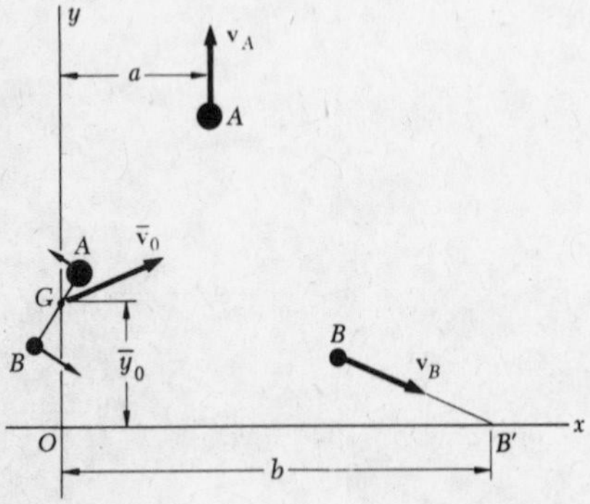

Fig. P14.28

14.28 Two small disks A and B, of mass 2 kg and 1 kg, respectively, may slide on a horizontal and frictionless surface. They are connected by a cord of negligible mass and spin about their mass center G. At $t = 0$, the coordinates of G are $\bar{x}_0 = 0$, $\bar{y}_0 = 1.6$ m, and its velocity is $\bar{\mathbf{v}}_0 = (1.5 \text{ m/s})\mathbf{i} + (1.2 \text{ m/s})\mathbf{j}$. Shortly thereafter, the cord breaks and disk A is observed to move along a path parallel to the y axis at a distance $a = 1.96$ m from that axis. Knowing that, initially, the angular momentum of the two disks about G was 3 kg · m²/s counterclockwise and that their kinetic energy relative to a centroidal frame was 18.75 J, determine (*a*) the velocities of A and B after the cord breaks, (*b*) the abscissa b of the point B' where the path of B intersects the x axis.

SECTIONS 14.10 to 14.12

Note. In the following problems use $\rho = 1000 \text{ kg/m}^3$ for the density of water in SI units, and $\gamma = 62.4 \text{ lb/ft}^3$ for its specific weight in U.S. customary units.

14.29 A stream of water of cross-sectional area A and velocity $\mathbf{v}_1$ strikes the curved surface of a block which is held motionless ($V = 0$) by the forces $\mathbf{P}_x$ and $\mathbf{P}_y$. Determine the magnitudes of $\mathbf{P}_x$ and $\mathbf{P}_y$ when $A = 500 \text{ mm}^2$ and $v_1 = 40 \text{ m/s}$.

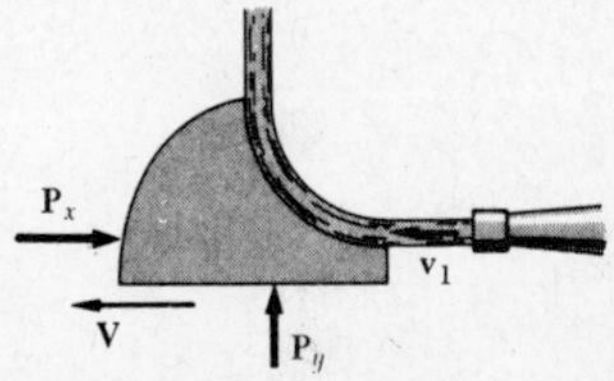

Fig. P14.29 and P14.30

14.30 A stream of water of cross-sectional area A and velocity $\mathbf{v}_1$ strikes the curved surface of a block which moves to the left with a velocity $\mathbf{V}$. Determine the magnitudes of the forces $\mathbf{P}_x$ and $\mathbf{P}_y$ required to hold the block when $A = 3 \text{ in}^2$, $v_1 = 90 \text{ ft/s}$, and $V = 25 \text{ ft/s}$.

14.31 Water flows in a continuous sheet from between two plates A and B with a velocity $\mathbf{v}$. The stream is split into two equal streams *1* and *2* by a vane attached to plate C. Denoting the total rate of flow by Q, determine the force exerted by the stream on plate C.

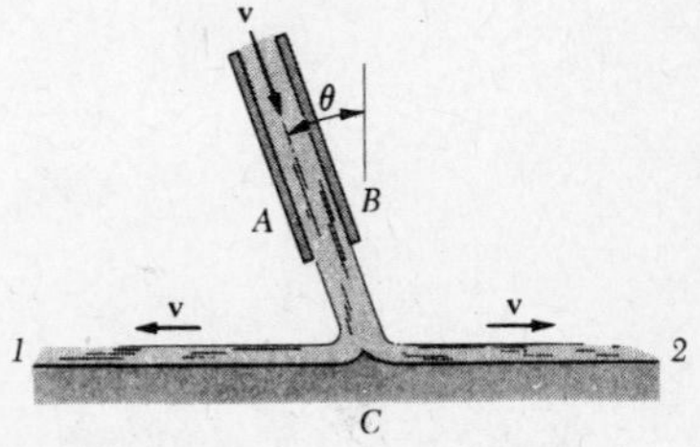

Fig. P14.31

14.32 Water flows in a continuous sheet from between two plates A and B with a velocity $\mathbf{v}$. The stream is split into two parts by a smooth horizontal plate C. Denoting the total rate of flow by Q, determine the rate of flow of each of the resulting streams. (*Hint.* The plate C can exert only a vertical force on the water.)

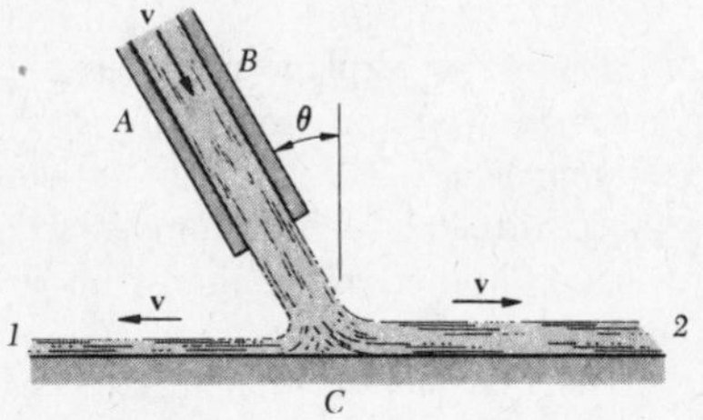

Fig. P14.32

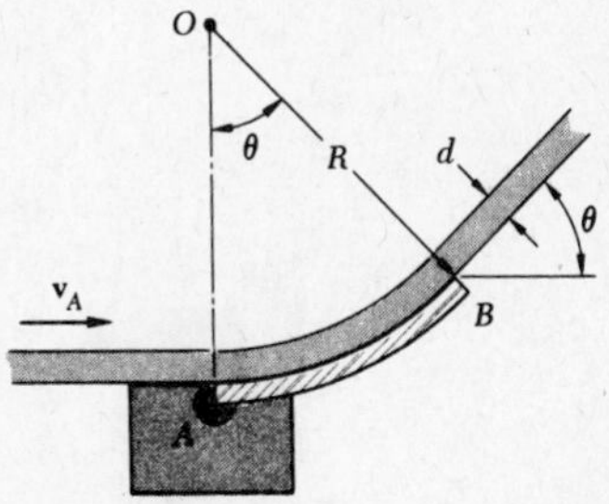

Fig. P14.33

14.33 A stream of water of cross-sectional area A and velocity $\mathbf{v}_A$ is deflected by a vane AB in the shape of an arc of circle of radius R. Knowing that the vane is welded to a fixed support at A, determine the components of the force-couple system exerted by the support on the vane.

14.34 The stream of water shown flows at the rate of $0.9 \text{ m}^3/\text{min}$ and moves with a velocity of magnitude 30 m/s at both A and B. The vane is supported by a pin connection at C and by a load cell at D which can exert only a horizontal force. Neglecting the weight of the vane, determine the reactions at C and D.

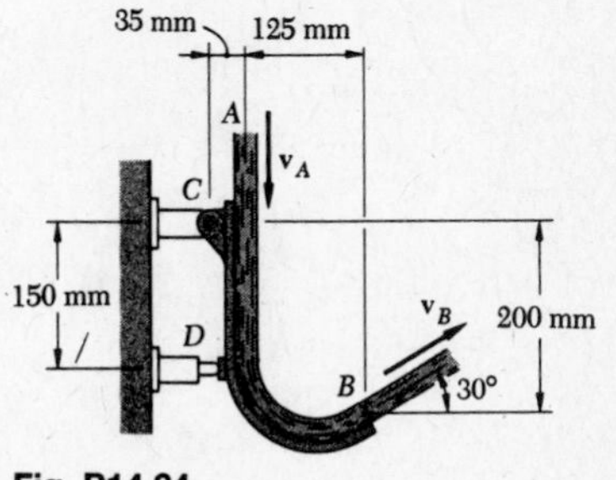

Fig. P14.34

14.35 A jet of water of cross-sectional area $A = 1.5$ in.^{2} moving with a downward velocity of magnitude $v_A = v_B = 60$ ft/sec is deflected by the two vanes shown which are welded to a rectangular plate. Knowing that the combined weight of the plate and vanes is $W = 20$ lb, determine the reactions at C and D.

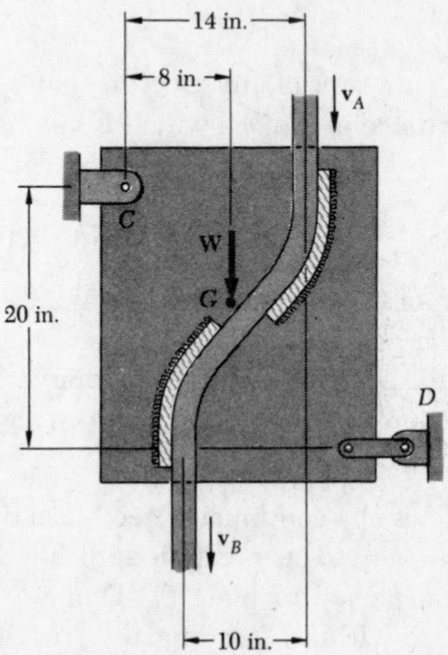

Fig. P14.35

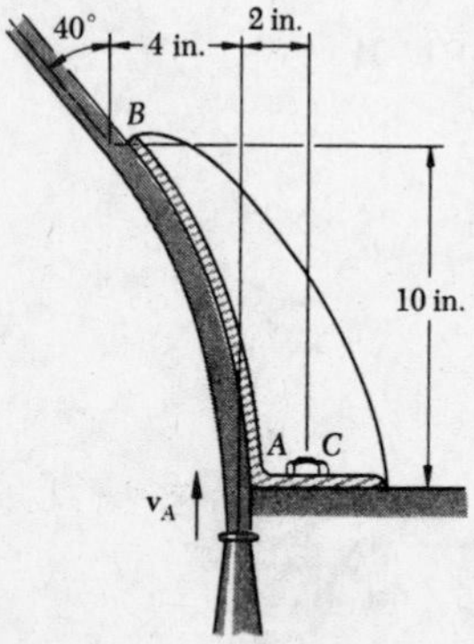

Fig. P14.36

14.36 The nozzle shown discharges 250 gal/min of water with a velocity $\mathbf{v}_A$ of 120 ft/s. The stream is deflected by the fixed vane AB. Determine the force-couple system which must be applied at C in order to hold the vane in place (1 ft^3 = 7.48 gal).

14.37 The propeller of an airplane produces a thrust of 4000 N when the airplane is at rest on the ground and has a slipstream of 2-m diameter. Assuming $\rho = 1.21$ kg/m^3 for air, determine (*a*) the speed of the air in the slipstream, (*b*) the volume of air passing through the propeller per second, (*c*) the kinetic energy imparted per second to the air of the slipstream.

14.38 While cruising in horizontal flight at a speed of 800 km/h, a 9000-kg jet airplane scoops in air at the rate of 70 kg/s and discharges it with a velocity of 600 m/s relative to the airplane. (*a*) Determine the total drag due to air friction. (*b*) Assuming that the drag is proportional to the square of the speed, determine the horizontal cruising speed if the flow of air through the jet is increased by 10 percent, i.e., to 77 kg/s.

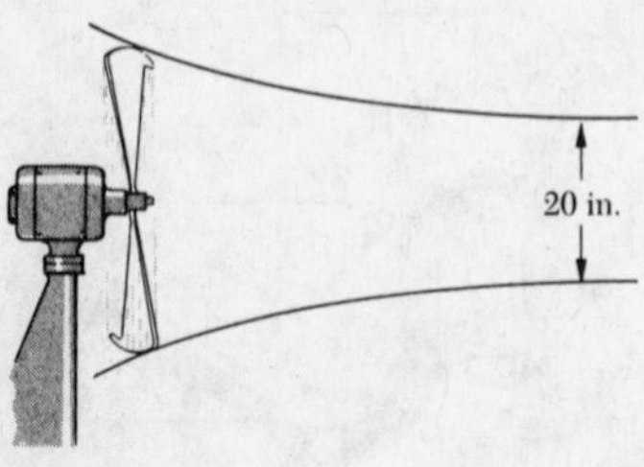

Fig. P14.39

14.39 The slipstream of a fan has a diameter of 20 in. and a velocity of 35 ft/sec relative to the fan. Assuming air weighs 0.076 lb/ft^3 and neglecting the velocity of approach of the air, determine the force required to hold the fan motionless.

14.40 An unloaded helicopter of weight 1800 lb produces a slipstream of 28-ft diameter. Assuming that air weighs 0.076 lb/ft^3, determine the vertical component of the velocity of the air in the slipstream when the helicopter is hovering in midair.

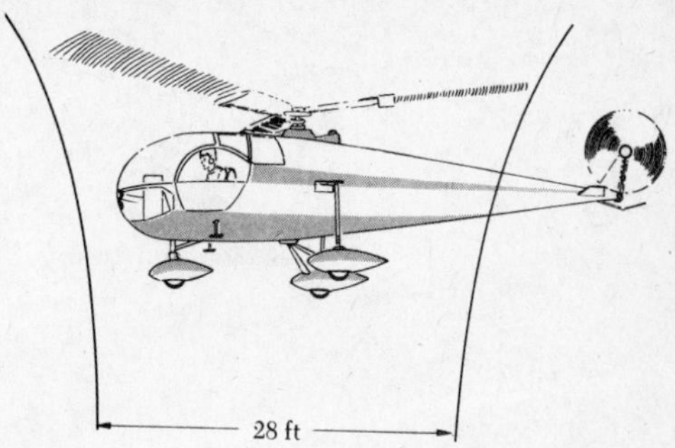

Fig. P14.40

14.41 Each arm of the sprinkler shown discharges water at the rate of 10 liters per minute with a velocity of 12 m/s relative to the arm. Neglecting the effect of friction, determine (*a*) the constant rate at which the sprinkler will rotate, (*b*) the couple **M** which must be applied to the sprinkler to hold it stationary.

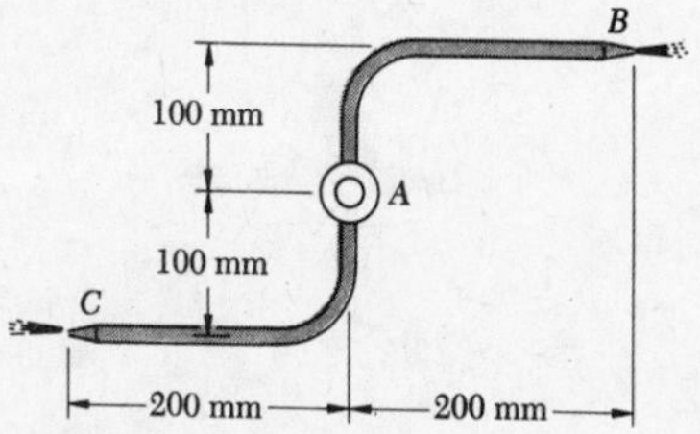

Fig. P14.41

14.42 For use in shallow water the pleasure boat shown is powered by a water jet. Water enters the engine through orifices located in the bow and is discharged through a horizontal pipe at the stern. Knowing that the water is discharged at the rate of 10 m^3/min with a velocity of 15 m/s relative to the boat, determine the propulsive force developed when the speed of the boat is (*a*) 6 m/s, (*b*) zero.

Fig. P14.42

14.43 The cruising speed of a jet airliner is 600 mi/h. Each of the four engines discharges air with a velocity of 2000 ft/s relative to the plane. Assuming that the drag due to air resistance is proportional to the square of the speed, determine the speed of the airliner when only two of the engines are in operation.

Fig. P14.43

14.44 The total drag due to air friction of a jet airplane traveling at 600 mph is 3500 lb. Knowing that the exhaust velocity is 2000 ft/sec relative to the airplane, determine the weight of air which must pass through the engine per second to maintain the speed of 600 mph in level flight.

14.45 Gravel falls with practically zero velocity onto a conveyor belt at the constant rate $q = dm/dt$. A force **P** is applied to the belt to maintain a constant speed v. Derive an expression for the angle θ for which the force **P** is zero.

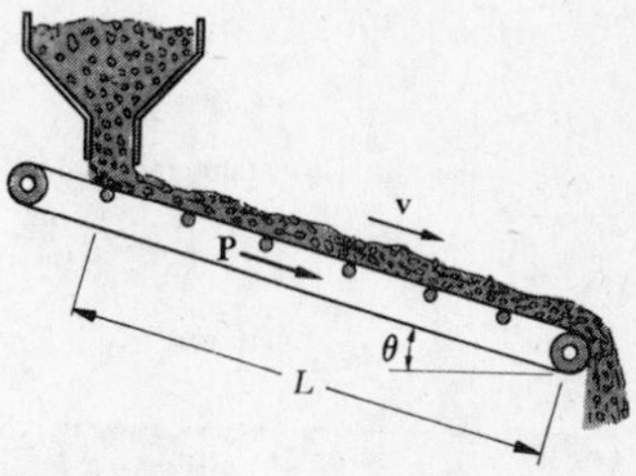

Fig. P14.45

14.46 Each of the two conveyor belts shown discharges sand at a constant rate of 2.25 kg/s. The sand falls through a height h and is deflected by a stationary vane. Knowing that the velocity of the sand is horizontal as it leaves the vane, determine the force **P** required to hold the vane when (*a*) $h = 2$ m, (*b*) $h = 4$ m.

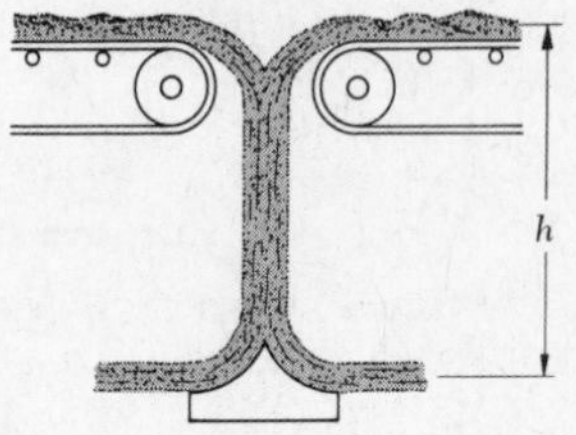

Fig. P14.46

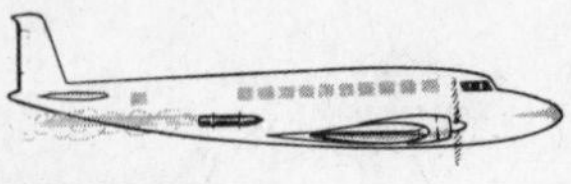

Fig. P14.47

14.47 A booster rocket is attached to the fuselage of an airplane weighing 40,000 lb. The rocket fuel is consumed at the rate of 10 lb/s and is ejected with a relative velocity of 5000 ft/s. Determine the additional propulsive thrust available while the rocket is being fired (*a*) if the speed of the airplane is 400 mi/h, (*b*) if the airplane is at rest on the ground.

14.48 Because of the play existing in the couplings of railroad freight cars, a long train may be started by successively setting each car in motion. If the engine moves at a constant speed v, (*a*) show that the magnitude of the constant force required to set the cars in motion is $F = v(dm/dt)$, where dm/dt is the rate at which the mass of the cars is set in motion. (*b*) Show that the work done in setting each car in motion is m_0v^2, where m_0 is the mass of a car. (*Note.* Although only half the work done is used to increase the kinetic energy of the train, this method of starting long trains is frequently used in order to overcome static friction.)

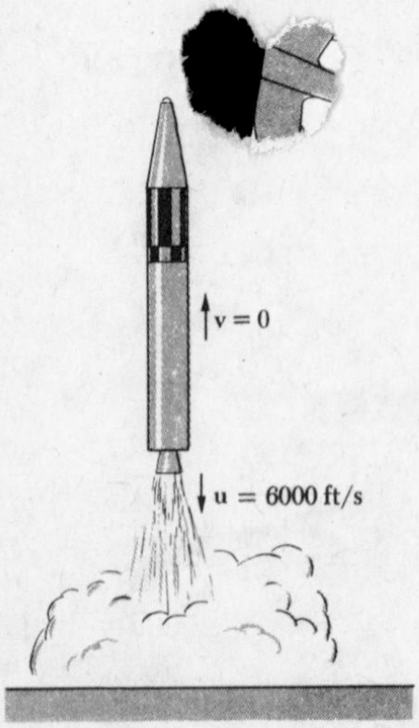

Fig. P14.49

14.49 A test rocket is designed to hover motionless above the ground. The shell of the rocket weighs 2500 lb, and the initial fuel load is 7500 lb. The fuel is burned and ejected with a velocity of 6000 ft/s. Determine the required rate of fuel consumption (*a*) when the rocket is fired, (*b*) as the last particle of fuel is being consumed.

14.50 An experimental space vehicle has a gross weight of 4000 lb, including 3000 lb of fuel. The fuel is consumed at the rate of 150 lb/s and is ejected with a relative velocity of 7000 ft/s. If the vehicle is fired vertically, determine its acceleration (*a*) as the engine is fired, (*b*) as the last particle of fuel is being consumed.

14.51 A rocket of gross mass 1000 kg, including 900 kg of fuel, is fired vertically when $t = 0$. Knowing that fuel is consumed at the rate of 10 kg/s and ejected with a relative velocity of 3500 m/s, determine the acceleration and velocity of the rocket when (*a*) $t = 0$, (*b*) $t = 45$ s, (*c*) $t = 90$ s.

14.52 The main engine installation of a space shuttle consists of three identical rocket engines which are required to provide a total thrust of 6000 kN. Knowing that the hydrogen-oxygen propellent is burned and ejected with a velocity of 3900 m/s, determine the required total rate of fuel consumption.

14.53 A spacecraft is launched vertically by a two-stage rocket. When the speed is 10,000 mi/h the first-stage-rocket casing is released and the second-stage rocket is fired. Fuel is consumed at the rate of 200 lb/s and ejected with a relative velocity of 8000 ft/s. Knowing that the combined weight of the second-stage rocket and spacecraft is 20,000 lb, including 17,000 lb of fuel, determine the maximum speed which can be attained by the spacecraft.

14.54 A multiple-stage rocket is more efficient than a rocket consisting of a single shell because expendable mass such as emptied fuel tanks can be discarded when no longer required. Assume that the rocket of Sample Prob. 14.8 is the second stage of a two-stage rocket and that, as the first section of the rocket is discarded, the velocity of the second section is $\mathbf{v}_0$ directed vertically upward. Taking $t = 0$ when the second stage begins, derive an expression for the velocity of the rocket at any time t.

14.55 A space tug describing a low-level circular orbit is to be transferred to a high-level orbit. The maneuver is started by firing the rocket engines to increase the speed of the tug from 7370 to 9850 m/s. The initial mass of the tug, fuel, and payload is 14.1 Mg. Knowing that the hydrogen-oxygen propellent is consumed at the rate of 20 kg/s and is ejected with a velocity of 3750 m/s, determine (*a*) the mass of fuel which must be expended to initiate the maneuver, (*b*) the time interval for which the engines must be fired.

14.56 Determine the distance between the spacecraft and the first-stage rocket casing of Prob. 14.53 as the last particle of fuel is being expelled by the second-stage rocket.

14.57 For the rocket of Sample Prob. 14.8, derive an expression for the height of the rocket as a function of the time t.

14.58 A space vehicle equipped with a retrorocket, which may expel fuel with a relative velocity $\mathbf{u}$, is moving with a velocity $\mathbf{v}_0$. Denoting by m_s the net mass of the vehicle and by m_f the mass of the unexpended fuel, determine the minimum ratio m_f/m_s for which the velocity of the vehicle can be reduced to zero.

CHAPTER 15
KINEMATICS OF RIGID BODIES

SECTIONS 15.1 to 15.4

15.1 The motion of a cam is defined by the relation $\theta = t^3 - 2t^2 - 4t + 10$, where θ is expressed in radians and t in seconds. Determine the angular coordinate, the angular velocity, and the angular acceleration of the cam when (*a*) $t = 0$, (*b*) $t = 3$ s.

15.2 For the cam of Prob. 15.1 determine (*a*) the time at which the angular velocity is zero, (*b*) the corresponding angular coordinate and angular acceleration.

15.3 The rotor of a steam turbine is rotating at a speed of 7200 rpm when the steam supply is suddenly cut off. It is observed that 5 min are required for the rotor to come to rest. Assuming uniformly accelerated motion, determine (*a*) the angular acceleration, (*b*) the total number of revolutions that the rotor executes before coming to rest.

15.4 The rotor of an electric motor has a speed of 1200 rpm when the power is cut off. The rotor is then observed to come to rest after executing 520 revolutions. Assuming uniformly accelerated motion, determine (*a*) the angular acceleration, (*b*) the time required for the rotor to come to rest.

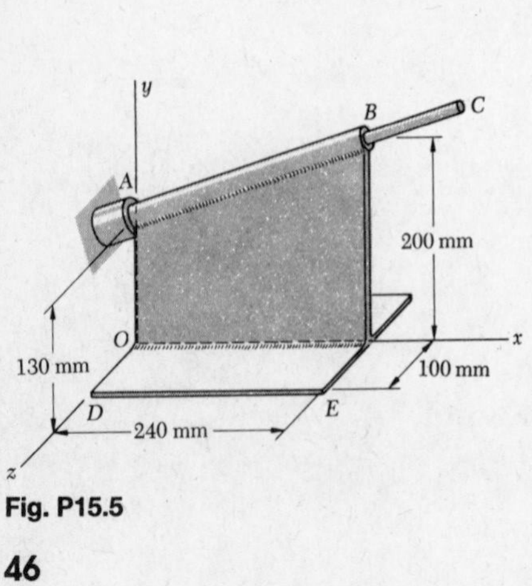

Fig. P15.5

15.5 The assembly shown rotates about the rod AC with a constant angular velocity of 5 rad/s. Knowing that at the instant considered, the velocity of corner D is downward, determine the velocity and acceleration of corner D.

15.6 In Prob. 15.5, determine the velocity and acceleration of corner E, assuming that the angular velocity is 5 rad/s and increases at the rate of 25 rad/s^2.

15.7 The rod $ABCD$ has been bent as shown and is supported by bearings at A and D. The rod rotates about the line joining points A and D with a constant angular velocity of 3 radians/sec. Knowing that, at the instant considered, the vertical component of the velocity of corner B is downward, determine the velocity and acceleration of corner C.

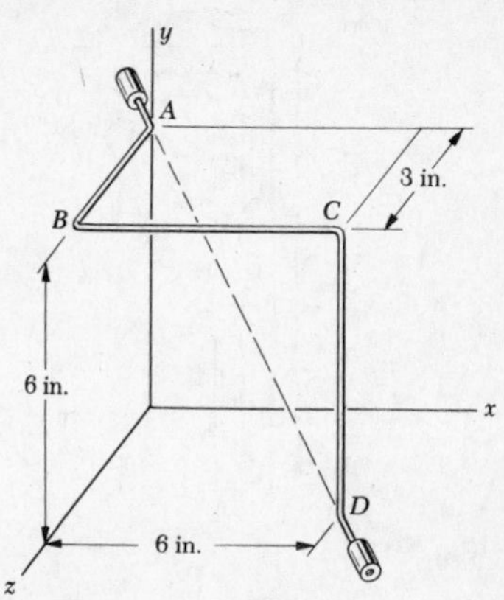

Fig. P15.7

15.8 Solve Prob. 15.7 assuming that the angluar velocity of 3 radians/sec about the line joining points A and D is being decreased at the rate of 6 radians/sec^2.

15.9 The friction wheel B executes 100 revolutions about its fixed shaft during the time interval t, while its angular velocity is being increased uniformly from 200 to 600 rpm. Knowing that wheel B rolls without slipping on the inside rim of wheel A, determine (*a*) the angular acceleration of wheel A, (*b*) the time interval t.

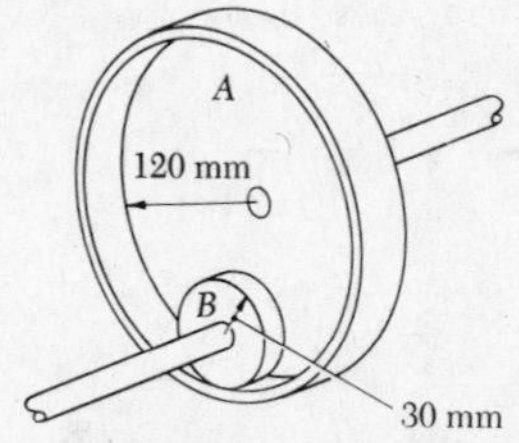

Fig. P15.9

15.10 Ring C has an inside diameter of 120 mm and hangs from the 40-mm-diameter shaft which rotates with a constant angular velocity of 30 rad/s. Knowing that no slipping occurs between the shaft and the ring, determine (*a*) the angular velocity of the ring, (*b*) the acceleration of the points of B and C which are in contact.

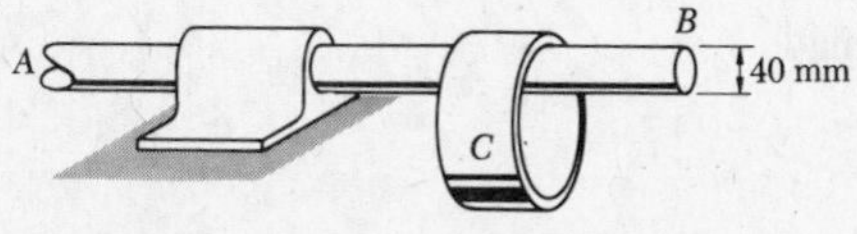

Fig. P15.10

15.11 At a given instant, the acceleration of the rack is 15 in./s^2 directed downward, and the velocity of the rack is 6 in./s directed upward. Determine (*a*) the angular acceleration and the angular velocity of the gear, (*b*) the total acceleration of the gear tooth in contact with the rack.

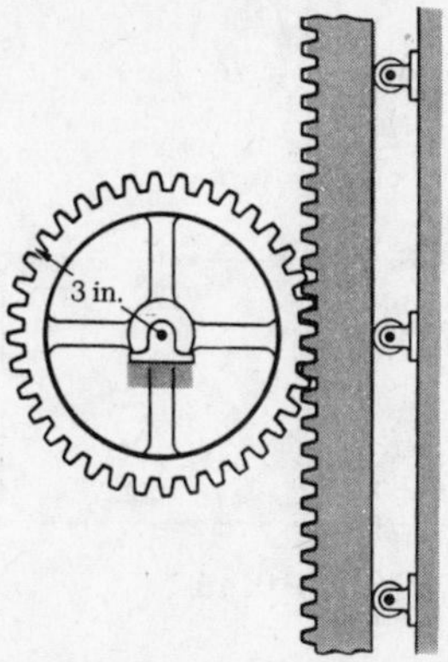

Fig. P15.11 and P15.12

15.12 The gear-and-rack system shown is initially at rest. If the acceleration of the rack is constant and equal to 0.5 ft/s^2 directed upward, determine (*a*) the angular velocity of the gear after the rack has moved 3 ft, (*b*) the time required for the gear to reach an angular velocity of 40 rpm.

15.13 A load is to be raised 6 m by the hoisting system shown. Assuming gear A is initially at rest, accelerates uniformly to a speed of 120 rpm in 5 s, and then maintains a constant speed of 120 rpm, determine (*a*) the number of revolutions executed by gear A in raising the load, (*b*) the time required to raise the load.

15.14 The system shown starts from rest at $t = 0$ and accelerates uniformly. Knowing that at $t = 4$ s the velocity of the load is 4.8 m/s downward, determine (*a*) the angular acceleration of gear A, (*b*) the number of revolutions executed by gear A during the 4-s interval.

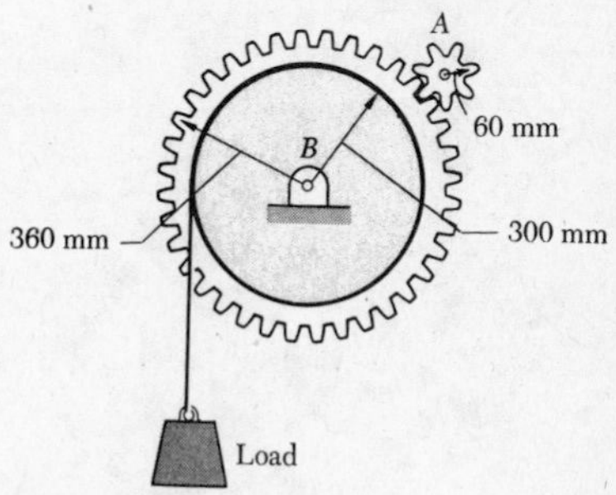

Fig. P15.13 and P15.14

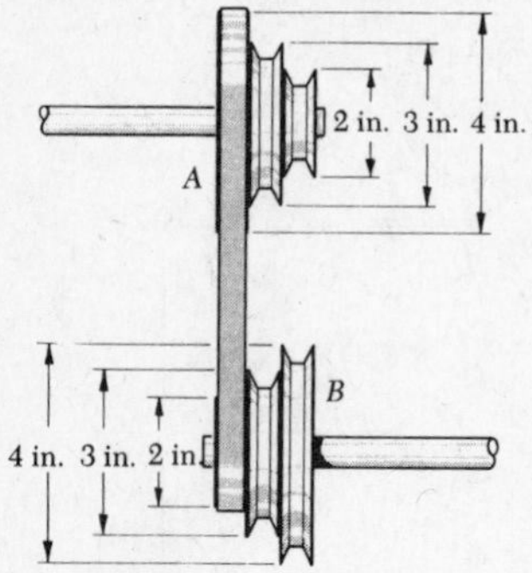

Fig. P15.15

15.15 The two pulleys shown may be operated with the V belt in any of three positions. If the angular acceleration of shaft A is 6 rad/s^2 and if the system is initially at rest, determine the time required for shaft B to reach a speed of 400 rpm with the belt in each of the three positions.

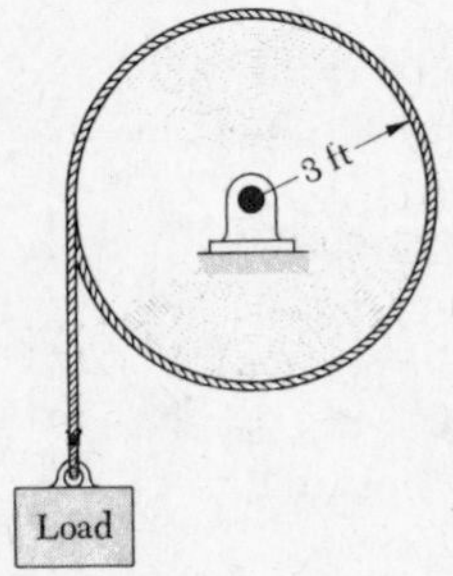

Fig. P15.16

15.16 The hoisting drum shown has an angular velocity of 120 rpm when the power is suddenly cut off. Knowing that the load rises through 60 more feet before coming to rest, determine (*a*) the angular acceleration of the drum, (*b*) the time required for the drum to come to rest.

15.17 A simple friction drive consists of two disks A and B. Initially, disk B has a clockwise angular velocity of 500 rpm, and disk A is at rest. It is known that disk B will coast to rest in 60 s. However, rather than waiting until both disks are at rest to bring them together, disk A is given a constant angular acceleration of 3 rad/s^2 counterclockwise. Determine (*a*) at what time the disks may be brought together if they are not to slip, (*b*) the angular velocity of each disk as contact is made.

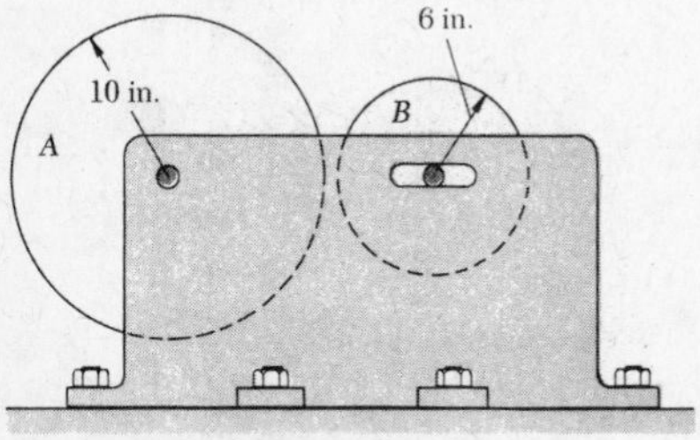

Fig. P15.17 and P15.18

15.18 Disk B is at rest when it is brought into contact with disk A which is rotating freely at 500 rpm clockwise. After 5 s of slippage, during which each disk has a constant angular acceleration, disk A reaches a final angular velocity of 300 rpm clockwise. Determine the angular acceleration of each disk during the period of slippage.

15.19 The two friction wheels A and B are to be brought together. Wheel A has an initial angular velocity of 600 rpm clockwise and will coast to rest in 40 s, while wheel B is initially at rest and is given a constant counterclockwise angular acceleration of 2 rad/s^2. Determine (*a*) at what time the wheels may be brought together if they are not to slip, (*b*) the angular velocity of each wheel as contact is made.

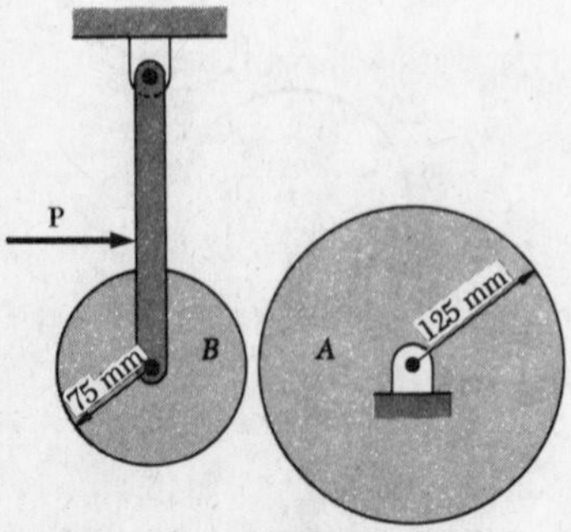

Fig. P15.19 and P15.20

15.20 Two friction wheels A and B are both rotating freely at 300 rpm clockwise when they are brought into contact. After 6 s of slippage, during which each wheel has a constant angular acceleration, wheel A reaches a final angular velocity of 60 rpm clockwise. Determine (*a*) the angular acceleration of each wheel during the period of slippage, (*b*) the time at which the angular velocity of wheel B is equal to zero.

SECTIONS 15.5 and 15.6

15.21 Rod AB is 6 ft long and slides with its ends in contact with the floor and the inclined plane. End A moves with a constant velocity of 20 ft/s to the right. At the instant when $\theta = 25°$, determine (*a*) the angular velocity of the rod, (*b*) the velocity of end B.

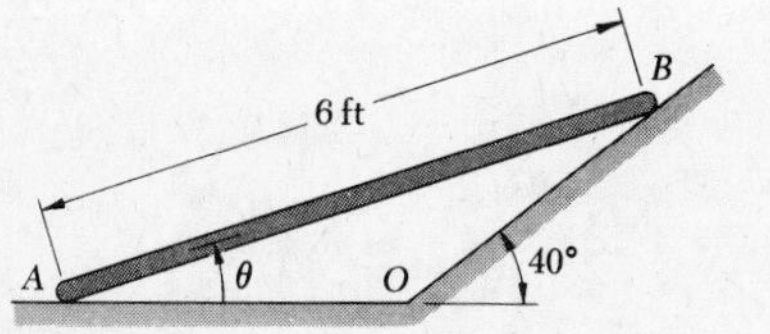

Fig. P15.21

15.22 Solve Prob. 15.21, assuming $\theta = 30°$.

15.23 Collar B moves with a constant velocity of 600 mm/s to the left. At the instant when $\theta = 30°$, determine (*a*) the angular velocity of rod AB, (*b*) the velocity of collar A.

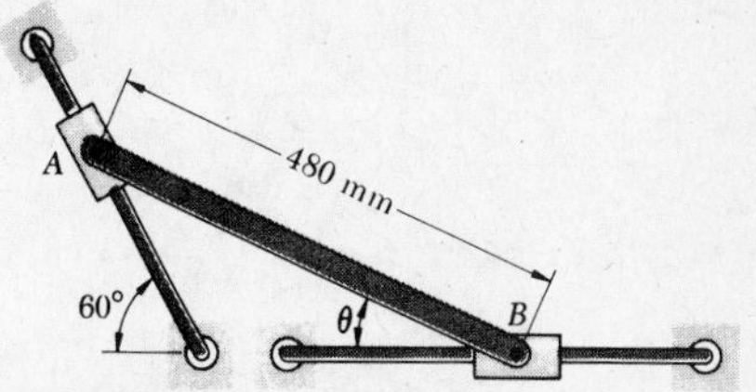

Fig. P15.23

15.24 Solve Prob. 15.23, assuming that $\theta = 45°$.

15.25 The rigid slab shown moves in the xy plane. Knowing that $(v_A)_x = 4$ in./sec, $(v_B)_y = -3$ in./sec, and $(v_C)_x = 16$ in./sec, determine (*a*) the angular velocity of the slab, (*b*) the velocity of point A.

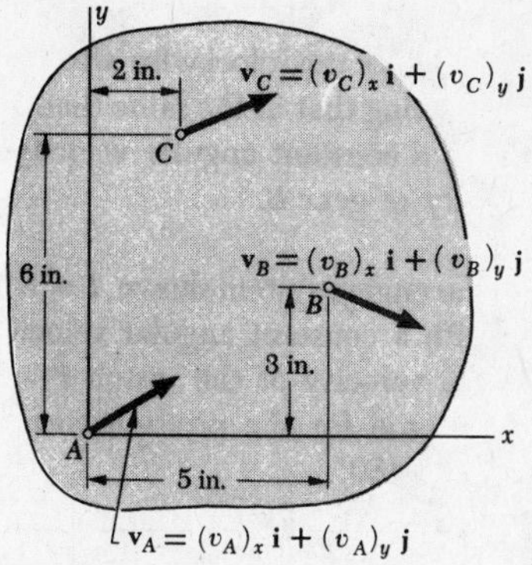

Fig. P15.25

15.26 In Prob. 15.25, determine the equation of the locus of the points of the slab for which the magnitude of the velocity is 6 in./sec.

15.27 In Prob. 15.25, determine (*a*) the velocity of point B, (*b*) the point of the slab with zero velocity.

15.28 The plate shown moves in the xy plane. Knowing that $(v_A)_x = 80$ mm/s, $(v_B)_y = 200$ mm/s, and $(v_C)_y = -40$ mm/s, determine (*a*) the angular velocity of the plate, (*b*) the velocity of point A.

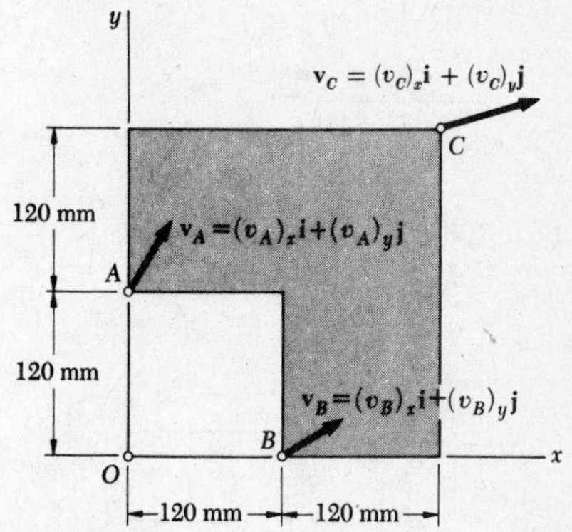

Fig. P15.28

15.29 In Prob. 15.28, determine the equation of the locus of the points of the plate for which the magnitude of the velocity is 100 mm/s.

15.30 In Prob. 15.28, determine (*a*) the velocity of point B, (*b*) the point of the plate with zero velocity.

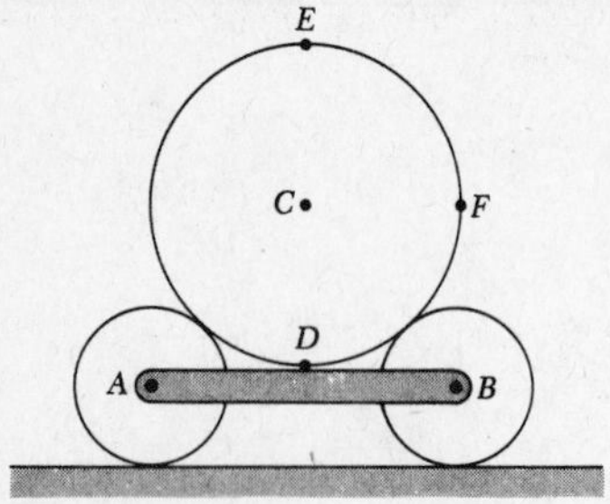

Fig. P15.31

15.31 Two rollers A and B of radius r are joined by a link AB and roll along a horizontal surface. A drum C of radius $2r$ is placed on the rollers as shown. If the link moves to the right with a constant velocity **v**, determine (*a*) the angular velocity of the rollers and of the drum, (*b*) the velocity of points D, E, and F of the drum.

15.32 Crank AB has a constant angular velocity of 12 rad/s clockwise. Determine the angular velocity of rod BD and the velocity of collar D when (*a*) $\theta = 0$, (*b*) $\theta = 90°$, (*c*) $\theta = 180°$.

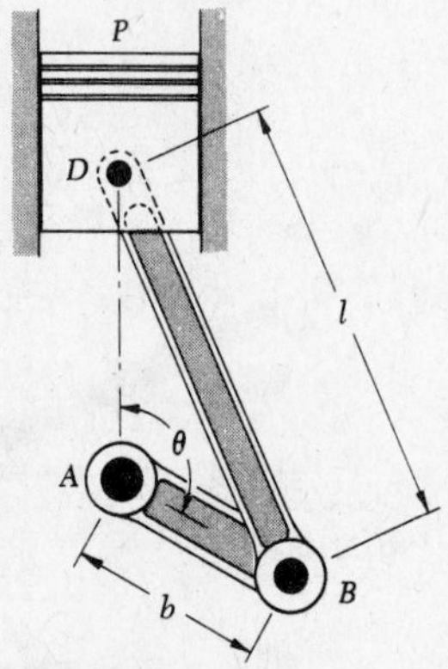

Fig. P15.33

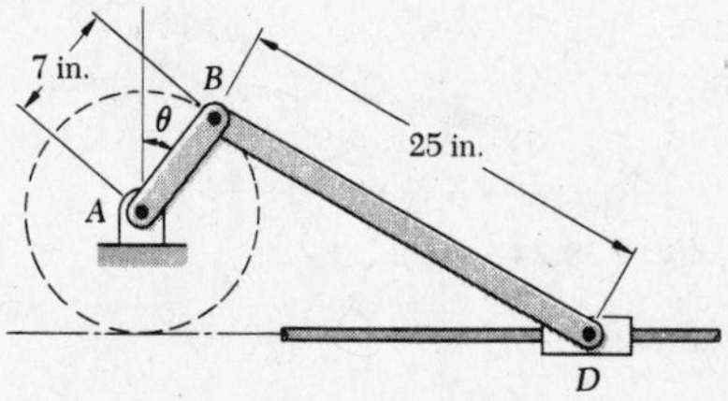

Fig. P15.32

15.33 In the engine system shown, $l = 160$ mm and $b = 60$ mm; the crank AB rotates with a constant angular velocity of 1000 rpm clockwise. Determine the velocity of the piston P and the angular velocity of the connecting rod for the position corresponding to (*a*) $\theta = 0$, (*b*) $\theta = 90°$, (*c*) $\theta = 180°$.

15.34 Solve Prob. 15.33 for the position corresponding to $\theta = 60°$.

15.35 Solve Prob. 15.32 for the position corresponding to $\theta = 120°$.

15.36 through 15.39 In the position shown, bar AB has a constant angular velocity of 3 rad/s counterclockwise. Determine the angular velocity of bars BD and DE.

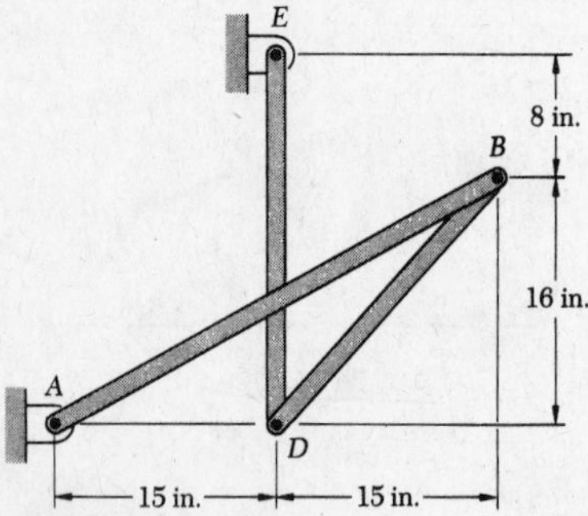

Fig. P15.36

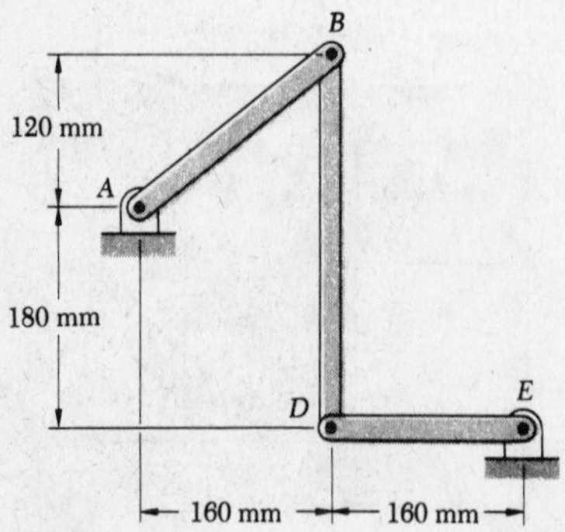

Fig. P15.37

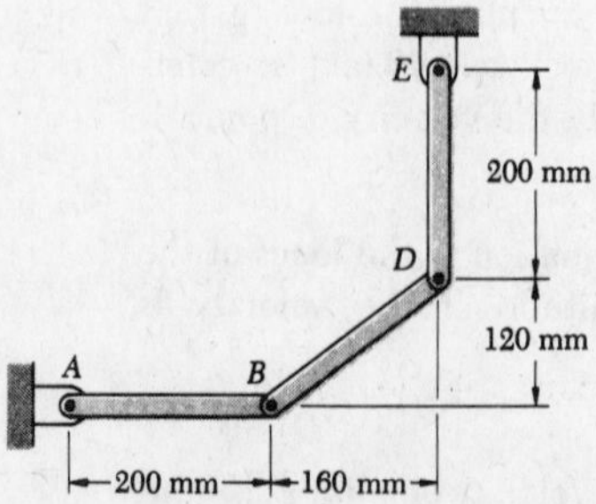

Fig. P15.38

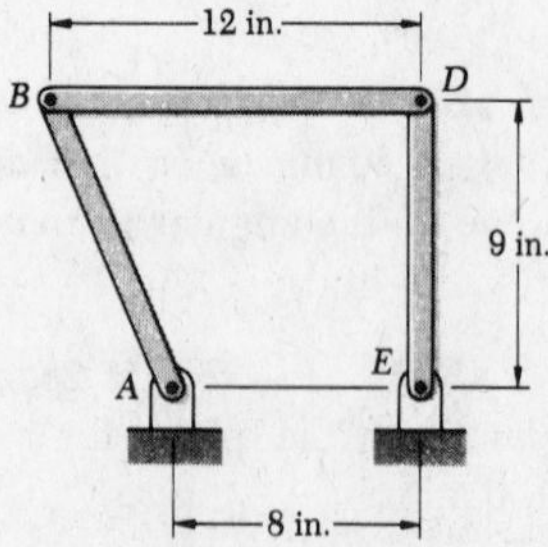

Fig. P15.39

SECTION 15.7

15.40 A helicopter moves horizontally in the x direction at a speed of 45 mi/h. Knowing that the main blades rotate clockwise at an angular velocity of 120 rpm, determine the instantaneous axis of rotation of the main blades.

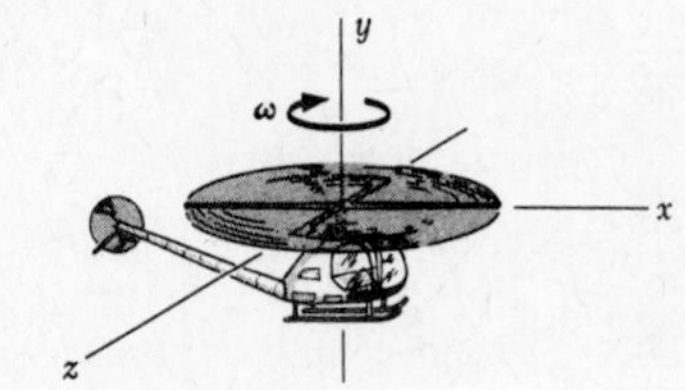

Fig. P15.40

15.41 A double pulley is attached to a slider block by a pin at A. The 2-in.-radius inner pulley is rigidly attached to the 4-in.-radius outer pulley. Knowing that each of the two cords is pulled at a constant speed of 12 in./sec as shown, determine (a) the instantaneous center of rotation of the double pulley, (b) the velocity of the slider block, (c) the number of inches of cord wrapped or unwrapped on each pulley per second.

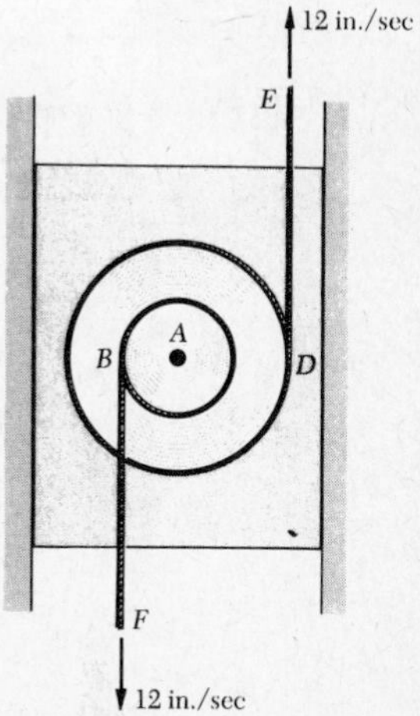

Fig. P15.41

15.42 A double pulley rolls without sliding on the plate AB, which moves to the left at a constant speed of 24 mm/s. The 60-mm-radius inner pulley is rigidly attached to the 80-mm-radius outer pulley. Knowing that cord E is pulled at a constant speed of 60 mm/s as shown, determine (a) the angular velocity of the pulley, (b) the velocity of the center G of the pulley.

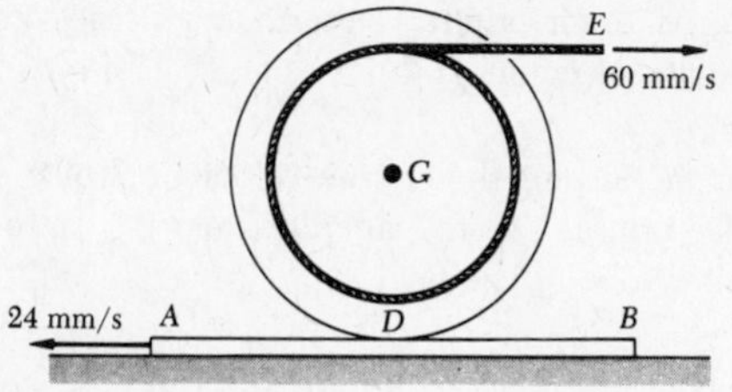

Fig. P15.42

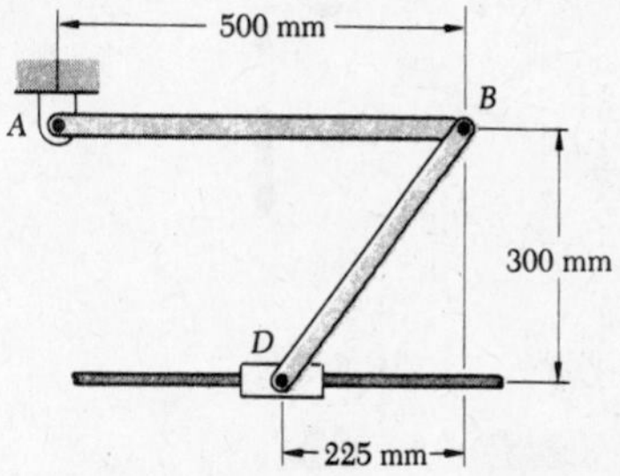

Fig. P15.43 and P15.44

15.43 Knowing that at the instant shown the angular velocity of crank AB is 3 rad/s clockwise, determine (a) the angular velocity of link BD, (b) the velocity of collar D, (c) the velocity of the midpoint of link BD.

15.44 Knowing that at the instant shown the velocity of collar D is 1.5 m/s to the right, determine (a) the angular velocities of crank AB and link BD, (b) the velocity of the midpoint of link BD.

15.45 The rod ABD is guided by wheels which roll in the tracks shown. Knowing that $\beta = 60°$ and that the velocity of A is 24 in./s downward, determine (a) the angular velocity of the rod, (b) the velocity of point D.

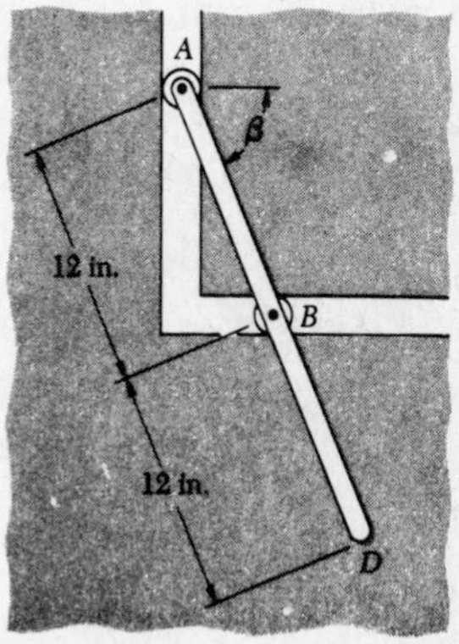

Fig. P15.45

15.46 Solve Prob. 15.45, assuming that $\beta = 30°$.

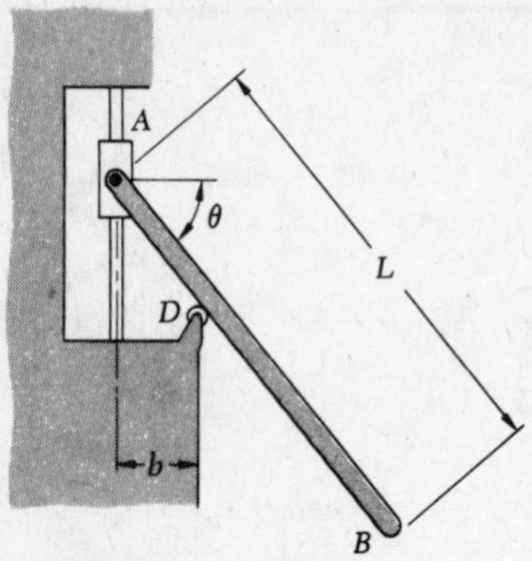

Fig. P15.47 and P15.48

15.47 Collar A slides downward with a constant velocity $\mathbf{v}_A$. Determine the angle θ corresponding to the position of rod AB for which the velocity of B is horizontal.

15.48 Collar A slides downward with a constant speed of 16 in./s. Knowing that $b = 2$ in., $L = 10$ in., and $\theta = 60°$, determine (*a*) the angular velocity of rod AB, (*b*) the velocity of B.

15.49 Two rods AB and BD are connected to three collars as shown. Knowing that collar A moves downward with a constant velocity of 120 mm/s, determine at the instant shown (*a*) the angular velocity of each rod, (*b*) the velocity of collar D.

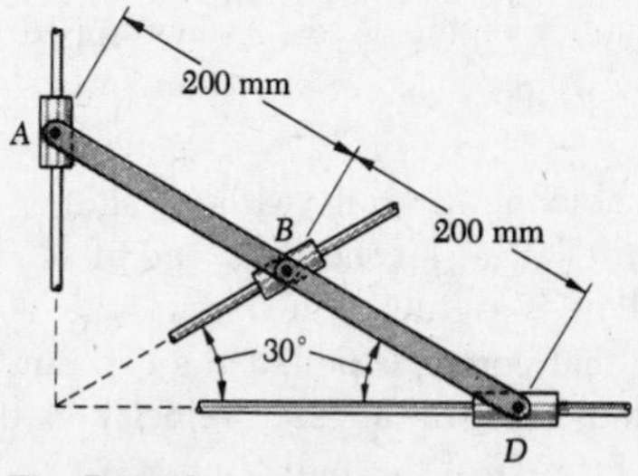

Fig. P15.49

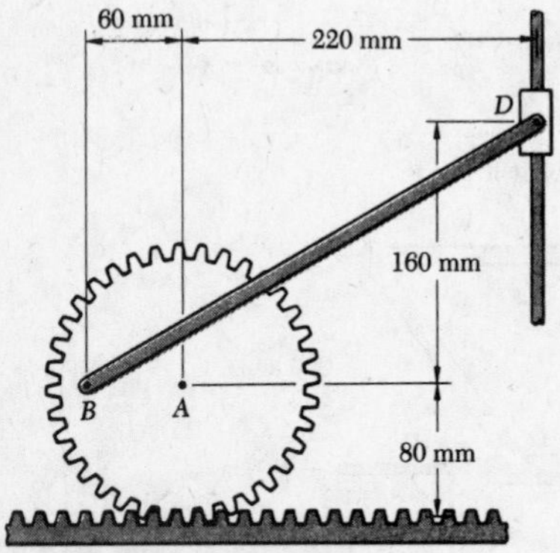

Fig. P15.50 and Fig. P15.51

15.50 At the instant shown, the velocity of the center of the gear is 200 mm/s to the right. Determine (*a*) the velocity of point B, (*b*) the velocity of collar D.

15.51 At the instant shown, the velocity of collar D is 360 mm/s downward. Determine (*a*) the angular velocity of rod BD, (*b*) the velocity of the center of the gear.

15.52 Describe the space centrode and the body centrode of gear A of Prob. 15.50 as the gear rolls on the horizontal rack.

15.53 Describe the space centrode and the body centrode of rod ABD of Prob. 15.45 as point A moves downward. (*Note:* The body centrode need not lie on a physical portion of the rod.)

15.54 Using the method of Sec. 15.7, solve Prob. 15.32.

15.55 Using the method of Sec. 15.7, solve Prob. 15.33.

15.56 Using the method of Sec. 15.7, solve Prob. 15.36.

15.57 Using the method of Sec. 15.7, solve Prob. 15.37.

15.58 Using the method of Sec. 15.7, solve Prob. 15.38

15.59 Using the method of Sec. 15.7, solve Prob. 15.39.

SECTIONS 15.8 and 15.9

15.60 A 16-ft steel beam is lowered by means of two cables unwinding at the same speed from overhead cranes. As the beam approaches the ground, the crane operators apply brakes to slow down the unwinding motion. The deceleration of the cable attached at A is $11\ \text{ft/s}^2$, while that of the cable attached at B is $3\ \text{ft/s}^2$. Determine the acceleration of point C and the angular acceleration of the beam at that instant.

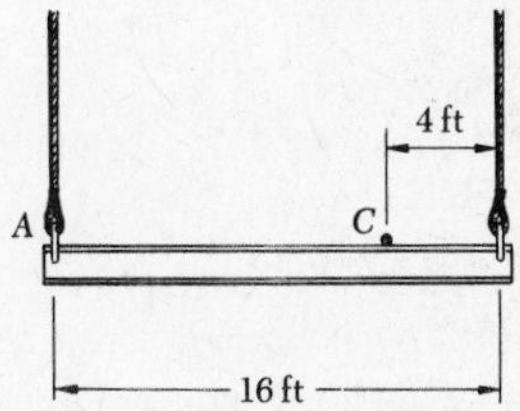

Fig. P15.60 and P15.61

15.61 The acceleration of point C is $2\ \text{ft/s}^2$ upward and the angular acceleration of the beam is $1.5\ \text{rad/s}^2$ clockwise. Knowing that the angular velocity of the beam is zero at the instant considered, determine the acceleration of each cable.

15.62 A 600-mm rod rests on a smooth horizontal table. A force **P** applied as shown produces the following accelerations: $\mathbf{a}_A = 0.8\ \text{m/s}^2$ to the right, $\alpha = 2\ \text{rad/s}^2$ clockwise as viewed from above. Determine the acceleration (*a*) of point B, (*b*) of point G.

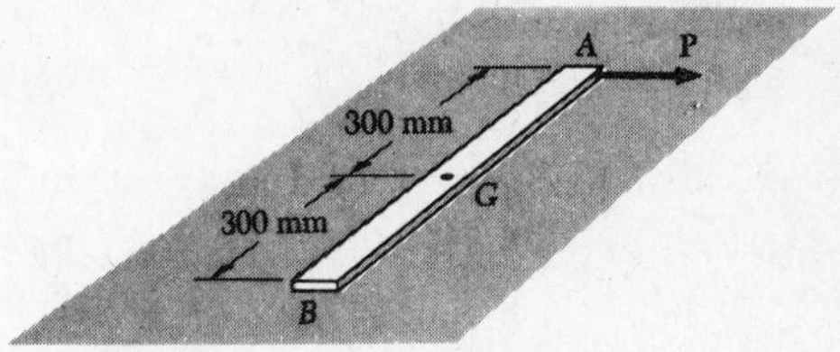

Fig. P15.62

15.63 In Prob. 15.62, determine the point of the rod which (*a*) has no acceleration, (*b*) has an acceleration of $0.350\ \text{m/s}^2$ to the right.

15.64 The moving carriage is supported by two casters A and C, each of $\frac{1}{2}$-in. diameter, and by a $\frac{1}{2}$-in.-diameter ball B. If at a given instant the velocity and acceleration of the carriage are as shown, determine (*a*) the angular accelerations of the ball and of each caster, (*b*) the accelerations of the center of the ball and of each caster.

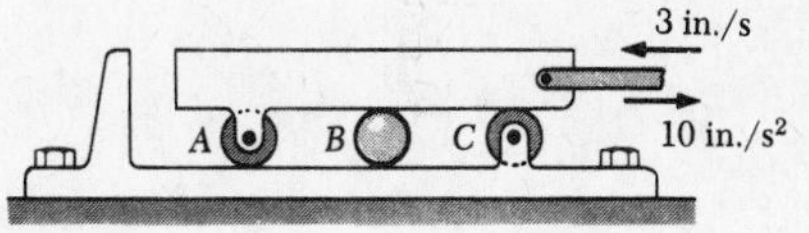

Fig. P15.64

15.65 and 15.66 The end A of the rod AB moves to the right with a constant velocity of 8 ft/s. For the position shown, determine (*a*) the angular acceleration of the rod, (*b*) the acceleration of the midpoint G of the rod.

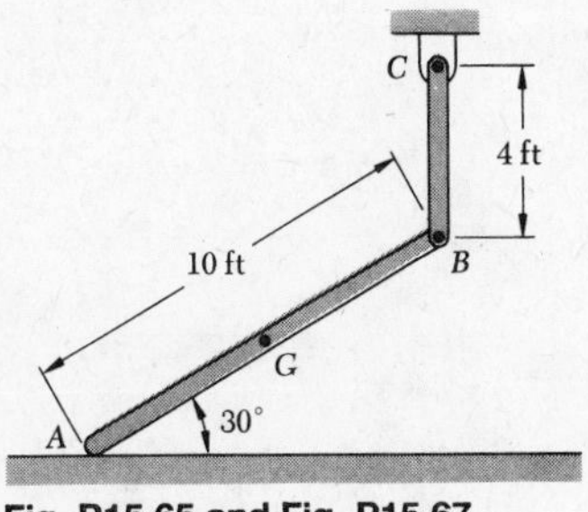

Fig. P15.65 and Fig. P15.67

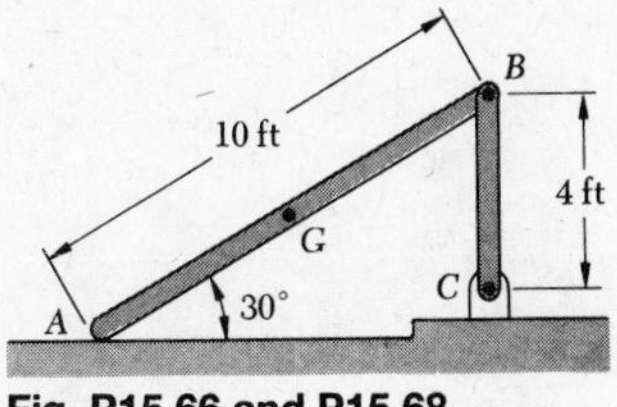

Fig. P15.66 and P15.68

15.67 and 15.68 In the position shown, end A of the rod AB has a velocity of 8 ft/s and an acceleration of $6\ \text{ft/s}^2$, both directed to the right. Determine (*a*) the angular acceleration of the rod, (*b*) the acceleration of the midpoint G of the rod.

15.69 and 15.70 The end A of the rod AB moves downward with a constant velocity of 300 mm/s. For the position shown, determine (*a*) the angular acceleration of the rod, (*b*) the acceleration of the midpoint G of the rod.

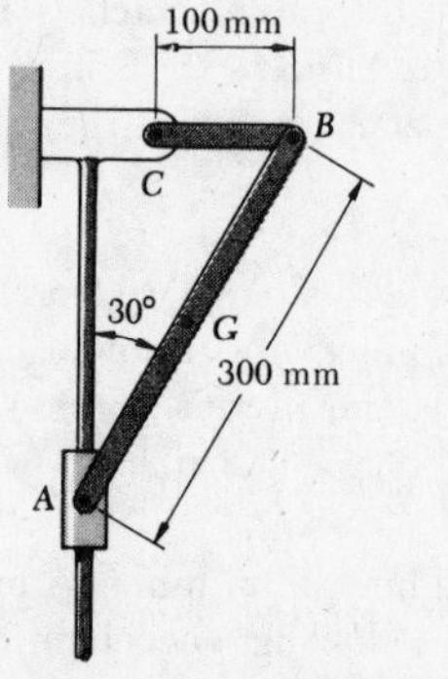

Fig. P15.69 and P15.71

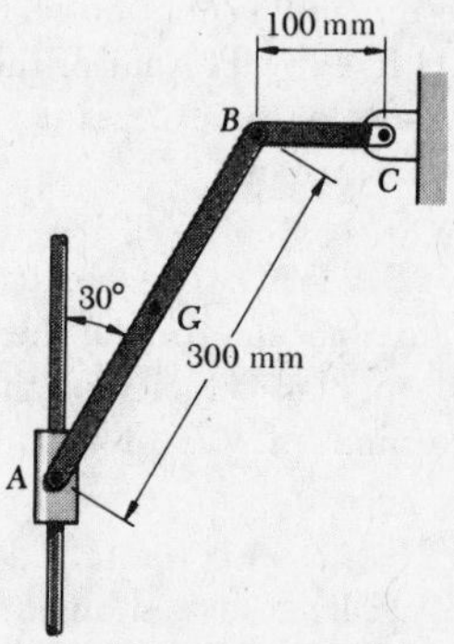

Fig. P15.70 and P15.72

15.71 and 15.72 In the position shown, end A of the rod AB has a velocity of 300 mm/s and an acceleration of 200 mm/s² both directed downward. Determine (*a*) the angular acceleration of the rod, (*b*) the acceleration of the midpoint G of the rod.

15.73 and 15.74 For the linkage indicated, determine the angular acceleration (*a*) of bar BD, (*b*) of bar DE.

15.73 Linkage of Prob. 15.37.
15.74 Linkage of Prob. 15.39.

15.75 and 15.76 At the instant shown, the disk rotates with a constant angular velocity ω_0 clockwise. Determine the angular velocities and the angular accelerations of the rods AB and BC.

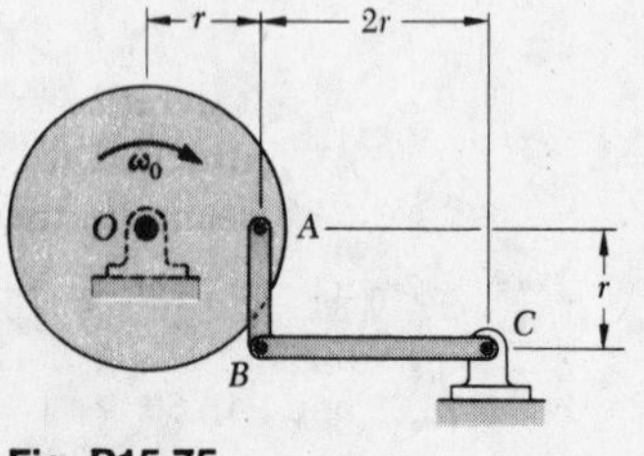

Fig. P15.75

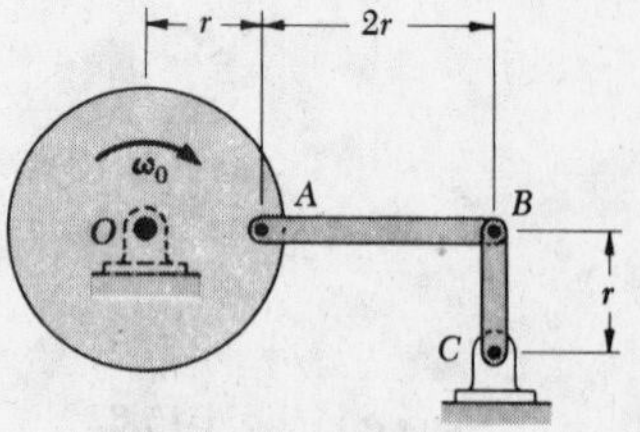

Fig. P15.76

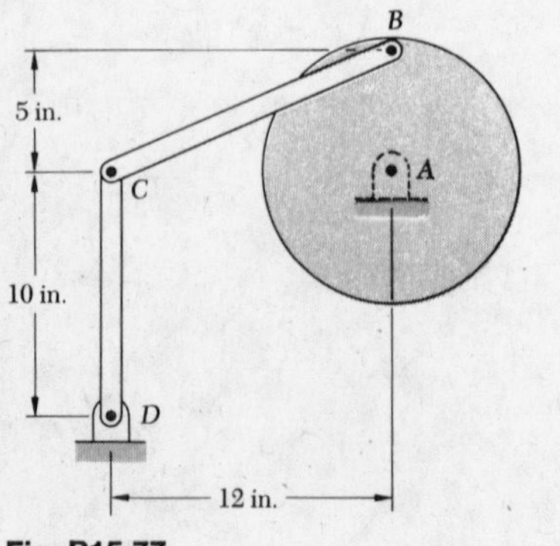

Fig. P15.77

15.77 The disk shown has a constant angular velocity of 6 rad/s clockwise. For the position shown, determine (*a*) the angular acceleration of each rod, (*b*) the acceleration of point C.

15.78 In the position shown, point A of bracket $ABCD$ has a velocity of magnitude $v_A = 250$ mm/s with $dv_A/dt = 0$. Determine (*a*) the angular acceleration of the bracket, (*b*) the acceleration of point C.

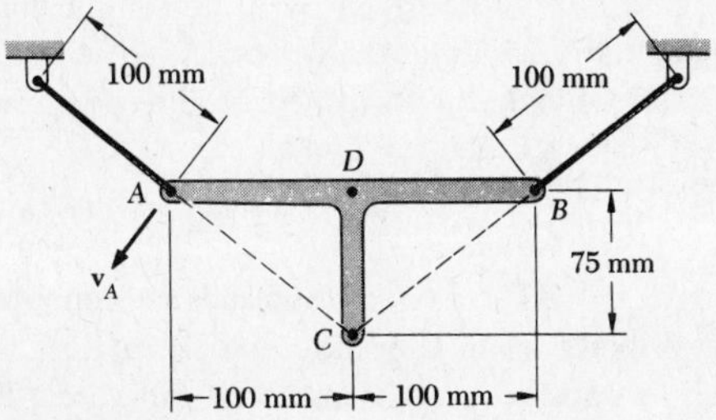

Fig. P15.78

15.79 In Prob. 15.78, determine the acceleration of point D.

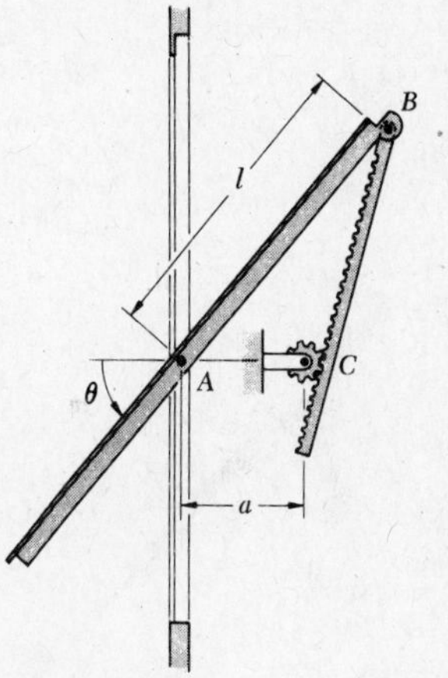

Fig. P15.80

***15.80** The position of a factory window is controlled by the rack and pinion shown. Knowing that the pinion C has a radius r and rotates counterclockwise at a constant rate ω, derive an expression for the angular velocity of the window.

***15.81** Knowing that rod AB rotates with an angular velocity $\boldsymbol{\omega}$ and an angular acceleration $\boldsymbol{\alpha}$, both counterclockwise, derive expressions for the components of the velocity and acceleration of point E.

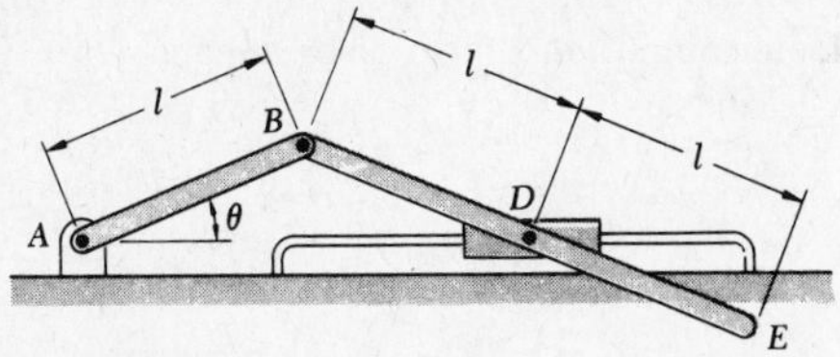

Fig. P15.81

***15.82** Collar B slides along rod OC and is attached to a sliding block which moves in a vertical slot. Knowing that rod OC rotates with an angular velocity $\boldsymbol{\omega}$ and with an angular acceleration $\boldsymbol{\alpha}$, both counterclockwise, derive expressions for the velocity and acceleration of collar B.

***15.83** Collar B slides along rod OC and is attached to a sliding block which moves upward with a constant velocity $\mathbf{v}$ in a vertical slot. Using the method of Sec. 15.9, derive an expression (*a*) for the angular velocity of rod OC, (*b*) for the angular acceleration of rod OC.

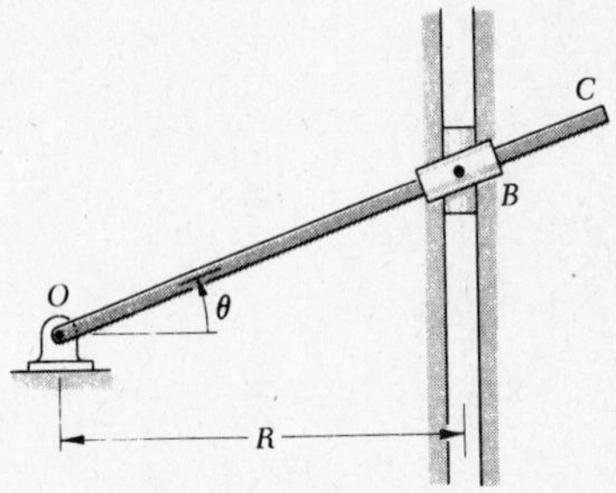

Fig. P15.82 and P15.83

SECTIONS 15.10 and 15.11

15.84 and 15.85 Two rotating rods are connected by a slider block P. The rod attached at B rotates with a constant clockwise angular velocity ω_B. For the given data, determine for the position shown (a) the angular velocity of the rod attached at A, (b) the relative velocity of the slider block P with respect to the rod on which it slides.

15.84 $b = 8$ in., $\omega_B = 6$ rad/s.

15.85 $b = 200$ mm, $\omega_B = 9$ rad/s.

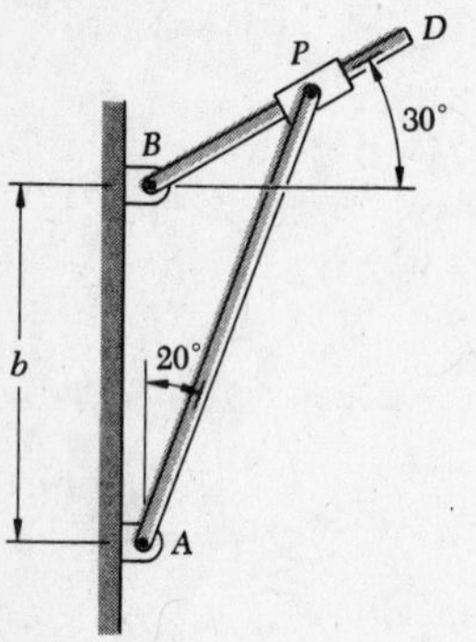

Fig. P15.84 and P15.86

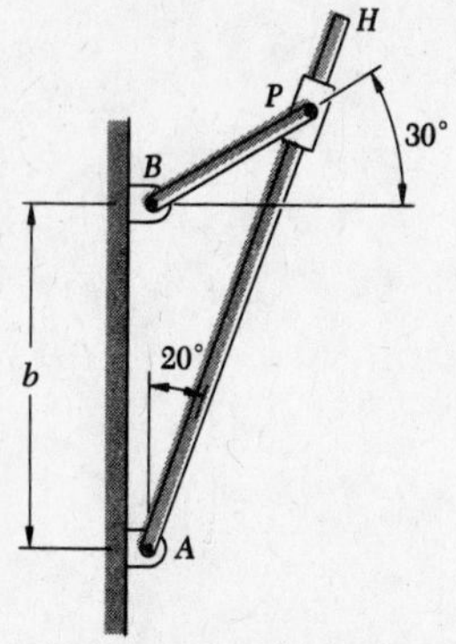

Fig. P15.85 and P15.87

15.86 and 15.87 Two rotating rods are connected by a slider block P. The velocity $\mathbf{v}_0$ of the slider block relative to the rod on which it slides is constant and is directed outward. For the given data, determine the angular velocity of each rod for the position shown.

15.86 $b = 200$ mm, $v_0 = 300$ mm/s.

15.87 $b = 8$ in., $v_0 = 12$ in./s.

15.88 Two rods AH and BD pass through smooth holes drilled in a hexagonal block. (The holes are drilled in different planes so that the rods will not hit each other.) Knowing that rod AH rotates counterclockwise at the rate ω, determine the angular velocity of rod BD and the relative velocity of the block with respect to each rod when (a) $\theta = 30°$, (b) $\theta = 15°$.

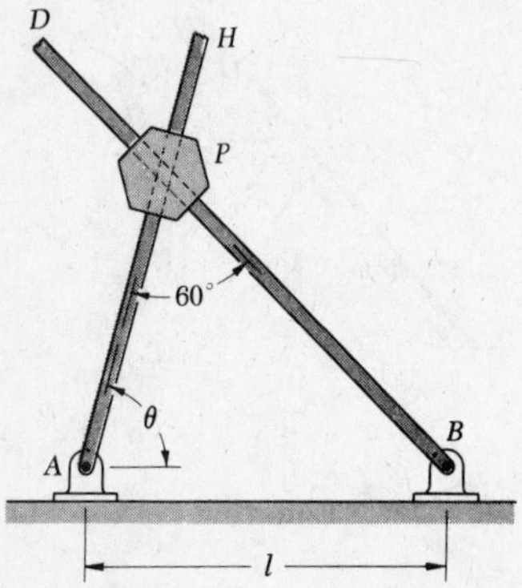

Fig. P15.88

15.89 Solve Prob. 15.88 when (a) $\theta = 90°$, (b) $\theta = 60°$.

15.90 A block P may slide on the arm OA which rotates at the constant angular velocity ω. As the arm rotates, the cord wraps around a *fixed* drum of radius b and pulls the block toward O with a speed $b\omega$. Determine the magnitude of the acceleration of the block in terms of r, b, and ω.

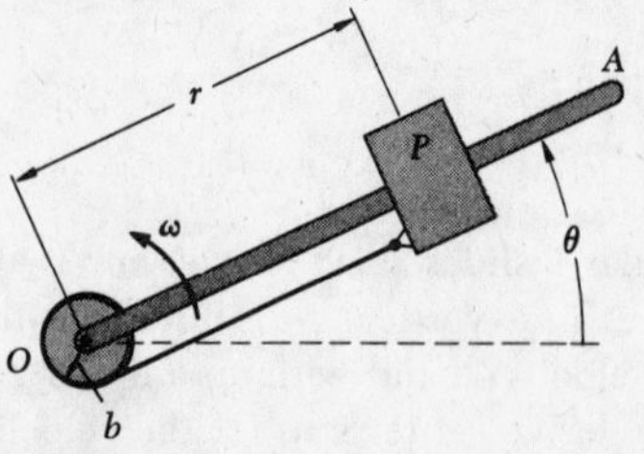

Fig. P15.90

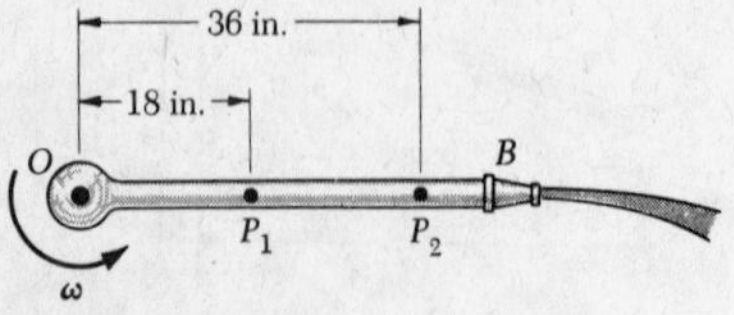

Fig. P15.91

15.91 Water flows through a straight pipe OB which rotates counterclockwise with an angular velocity of 120 rpm. If the velocity of the water relative to the pipe is 20 ft/s, determine the total acceleration (a) of the particle of water P_1, (b) of the particle of water P_2.

15.92 The cage of a mine elevator moves downward with a constant speed of 40 ft/s. Determine the magnitude and direction of the Coriolis acceleration of the cage if the elevator is located (*a*) at the equator, (*b*) at latitude 40° north, (*c*) at latitude 40° south. (*Hint.* In parts *b* and *c* consider separately the components of the motion parallel and perpendicular to the plane of the equator.)

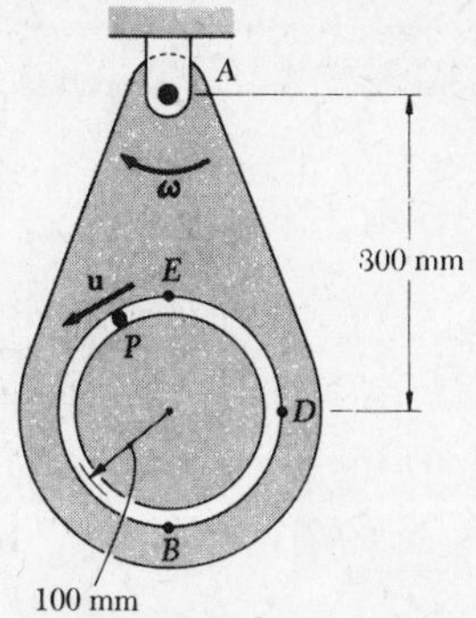

Fig. P15.94

15.93 A train crosses the parallel 50° north, traveling due north at a constant speed v. Determine the speed of the train if the Coriolis component of its acceleration is 3 mm/s². (See hint of Prob. 15.92.)

15.94 Pin P slides in the circular slot cut in the plate $ABDE$ at a constant relative speed $u = 0.5$ m/s as the plate rotates about A at the constant rate $\omega = 6$ rad/s. Determine the acceleration of the pin as it passes through (*a*) point B, (*b*) point D, (*c*) point E.

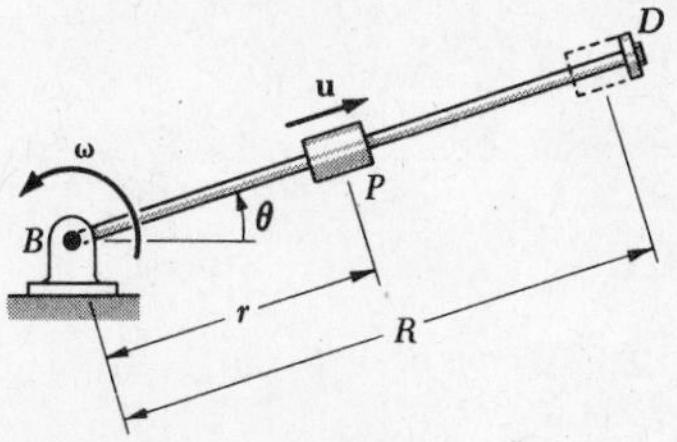

Fig. P15.96 and P15.97

15.95 Solve Prob. 15.94, assuming that at the instant considered the angular velocity $\boldsymbol{\omega}$ is being decreased at the rate of 10 rad/s² and that the relative velocity **u** is being decreased at the rate of 3 m/s².

15.96 The collar P slides outward at a constant relative speed u along the rod BD, which rotates at the constant angular velocity ω. Determine the magnitude of the acceleration of the collar P just before it reaches the end of the rod.

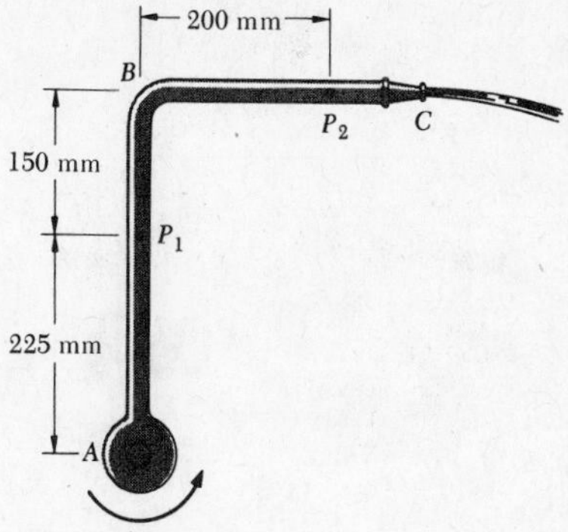

Fig. P15.98

15.97 The collar P slides outward at a constant relative speed u along the rod BD, which rotates at the constant angular velocity ω. Knowing that $r = 0$ when $\theta = 0$ and that the collar reaches D when $\theta = \pi$, determine the magnitude of the acceleration of the collar P just before it reaches the end of the rod.

15.98 Water flows through the sprinkler arm ABC with a velocity of 4.8 m/s relative to the arm. Knowing that the angular velocity of the arm is 90 rpm counterclockwise, determine at the instant shown the total acceleration (*a*) of the particle of water P_1, (*b*) of the particle of water P_2.

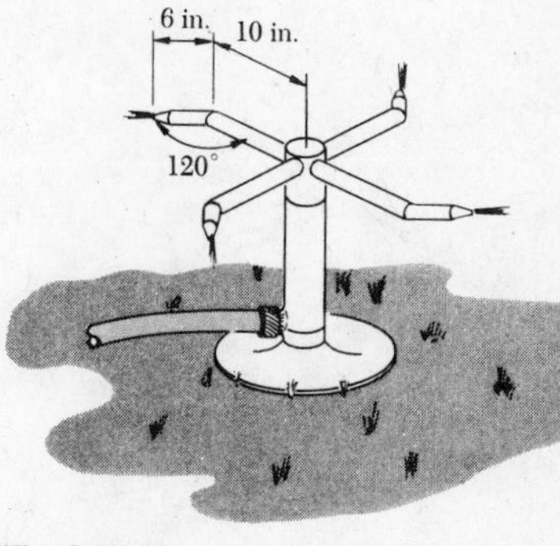

Fig. P15.99

15.99 A garden sprinkler has four rotating arms, each of which consists of two horizontal straight sections of pipe forming an angle of 120°. The sprinkler when operating rotates with a constant angular velocity of 180 rpm. If the velocity of the water relative to the pipe sections is 12 ft/s, determine the magnitude of the total acceleration of a particle of water as it passes the midpoint of (*a*) the 10-in. section of pipe, (*b*) the 6-in. section of pipe.

SECTIONS 15.12 and 15.13

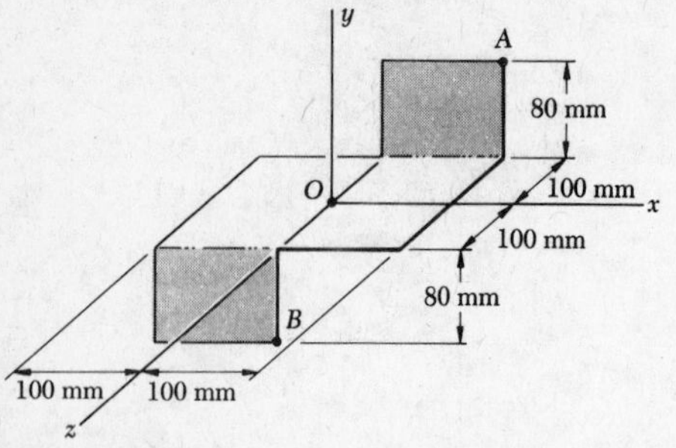

Fig. P15.100 and P15.101

15.100 The rigid body shown rotates about the origin of coordinates with an angular velocity $\boldsymbol{\omega}$. Denoting the velocity of point A by $\mathbf{v}_A = (v_A)_x\mathbf{i} + (v_A)_y\mathbf{j} + (v_A)_z\mathbf{k}$, and knowing that $(v_A)_x = 40$ mm/s and $(v_A)_y = -200$ mm/s, determine the velocity component $(v_A)_z$.

15.101 The rigid body shown rotates about the origin of coordinates with an angular velocity $\boldsymbol{\omega} = \omega_x\mathbf{i} + \omega_y\mathbf{j} + \omega_z\mathbf{k}$. Knowing that $(v_A)_y = 400$ mm/s, $(v_B)_y = -300$ mm/s, and $\omega_y = 2$ rad/s, determine (*a*) the angular velocity of the body, (*b*) the velocities of points A and B.

15.102 The circular plate and rod are rigidly connected and rotate about the ball-and-socket joint O with an angular velocity $\boldsymbol{\omega} = \omega_x\mathbf{i} + \omega_y\mathbf{j} + \omega_z\mathbf{k}$. Knowing that $\mathbf{v}_A = -(27 \text{ in./s})\mathbf{i} + (18 \text{ in./s})\mathbf{j} + (v_A)_z\mathbf{k}$ and $\omega_y = 4$ rad/s, determine (*a*) the angular velocity of the assembly, (*b*) the velocity of point B.

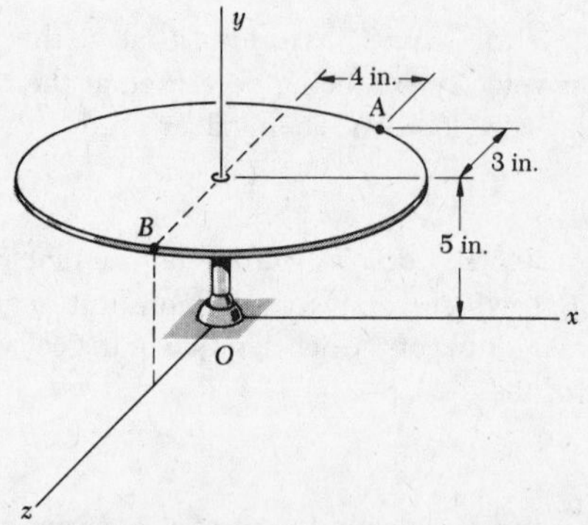

Fig. P15.102

15.103 Solve Prob. 15.102, assuming that $\omega_y = 0$.

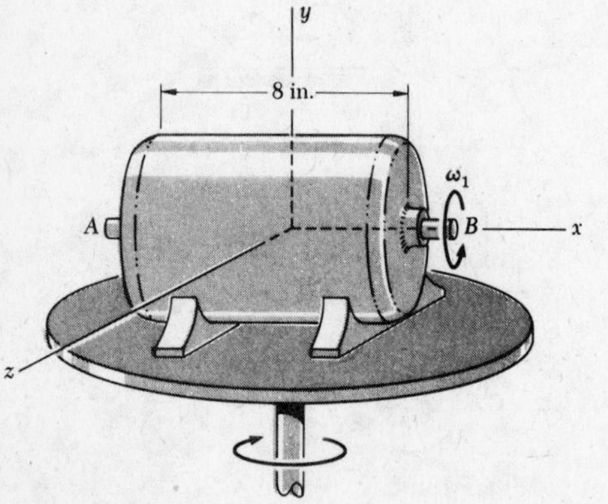

Fig. P15.104

15.104 The rotor of an electric motor rotates at the constant rate $\omega_1 = 3600$ rpm. Determine the angular acceleration of the rotor as the motor is rotated about the y axis with a constant angular velocity of 6 rpm clockwise when viewed from the positive y axis.

15.105 The propeller of a small airplane rotates at a constant rate of 2200 rpm in a clockwise sense when viewed by the pilot. Knowing that the airplane is turning left along a horizontal circular path of radius 1000 ft, and that the speed of the airplane is 150 mi/h, determine the angular acceleration of the propeller at the instant the airplane is moving due south.

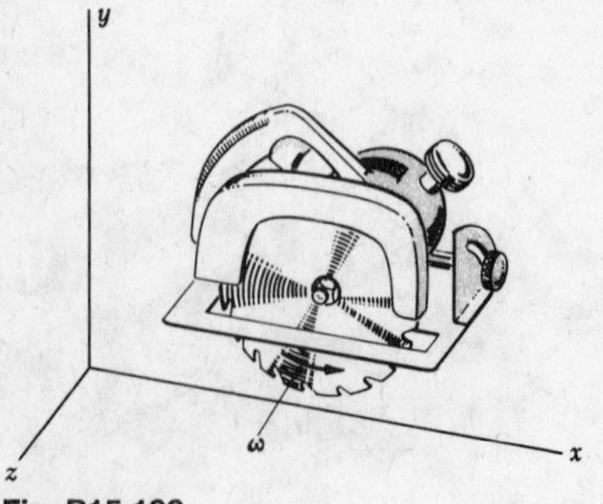

Fig. P15.106

15.106 The blade of a portable saw rotates at a constant rate $\omega = 1800$ rpm as shown. Determine the angular acceleration of the blade as a man rotates the saw about the y axis with an angular velocity of 3 rad/s and an angular acceleration of 5 rad/s^2, both clockwise when viewed from above.

15.107 Knowing that the turbine rotor shown rotates at a constant rate $\omega_1 = 10\,000$ rpm, determine the angular acceleration of the rotor if the turbine housing has a constant angular velocity of 3 rad/s clockwise as viewed from (*a*) the positive y axis, (*b*) the positive z axis.

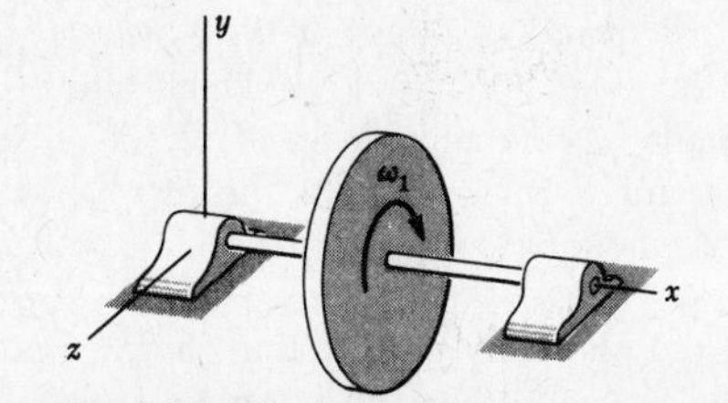

Fig. P15.107

15.108 In the gear system shown, gear A is free to rotate about the horizontal rod OA. Assuming that gear B is fixed and that shaft OC rotates with a constant angular velocity ω_1, determine (*a*) the angular velocity of gear A, (*b*) the angular acceleration of gear A.

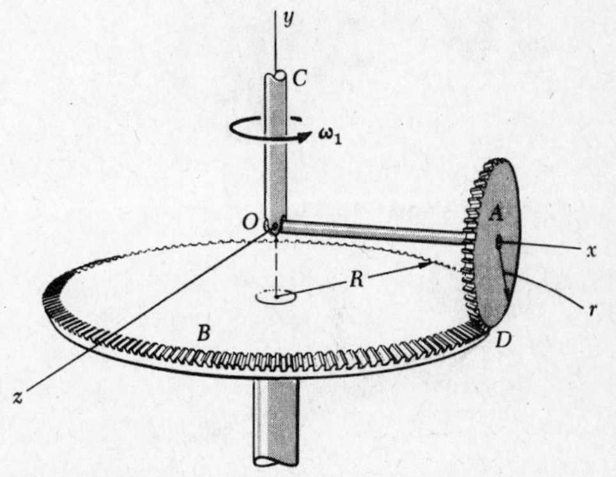

Fig. P15.108

15.109 Solve Prob. 15.108, assuming that shaft OC and gear B rotate with constant angular velocities ω_1 and ω_2, respectively, both counterclockwise as viewed from the positive y axis.

15.110 The radar antenna shown rotates with a constant angular velocity ω_1 of 1.5 rad/s about the y axis. At the instant shown the antenna is also rotating about the z axis with an angular velocity ω_2 of 2 rad/s and an angular acceleration $\boldsymbol{\alpha}_2$ of 2.5 rad/s^2. Determine (*a*) the angular acceleration of the antenna, (*b*) the accelerations of points A and B.

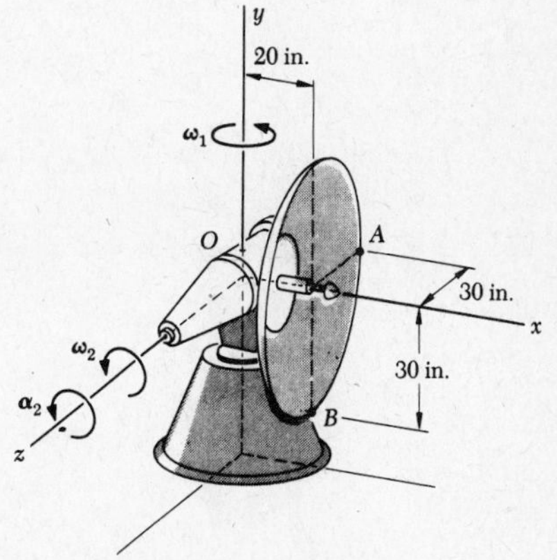

Fig. P15.110

15.111 A rod of length $OP = 500$ mm is mounted on a bracket as shown. At the instant considered the angle β is being increased at the constant rate $d\beta/dt = 4$ rad/s and the elevation angle γ is being increased at the constant rate $d\gamma/dt = 1.6$ rad/s. For the position $\beta = 0$ and $\gamma = 30°$, determine (*a*) the angular velocity of the rod, (*b*) the angular acceleration of the rod, (*c*) the velocity and acceleration of point P.

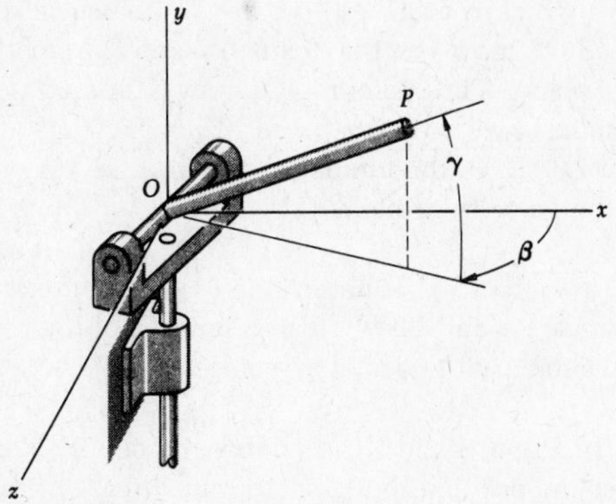

Fig. P15.111

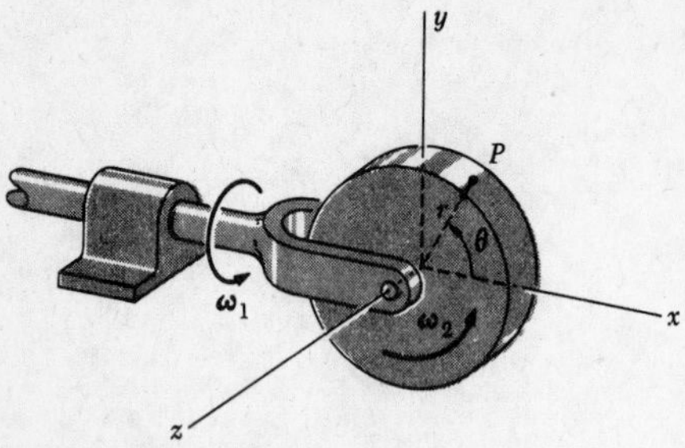

Fig. P15.112 and P15.113

15.112 A disk of radius r spins at the constant rate ω_2 about an axle held by a fork-ended horizontal rod which rotates at the constant rate ω_1. Determine the acceleration of point P for an arbitrary value of the angle θ.

15.113 A disk of radius r spins at the constant rate ω_2 about an axle held by a fork-ended horizontal rod which rotates at the constant rate ω_1. Determine (*a*) the angular acceleration of the disk, (*b*) the acceleration of point P on the rim of the disk when $\theta = 0$, (*c*) the acceleration of P when $\theta = 90°$.

15.114 In the planetary gear system shown, gears A and B are rigidly connected to each other and rotate as a unit about shaft FG. Gears C and D rotate with constant angular velocities of 15 rad/s and 30 rad/s, respectively (both counterclockwise when viewed from the right). Choosing the x axis to the right, the y axis upward, and the z axis pointing out of the plane of the figure, determine (*a*) the common angular velocity of gears A and B, (*b*) the angular velocity of shaft FH, which is rigidly attached to FG.

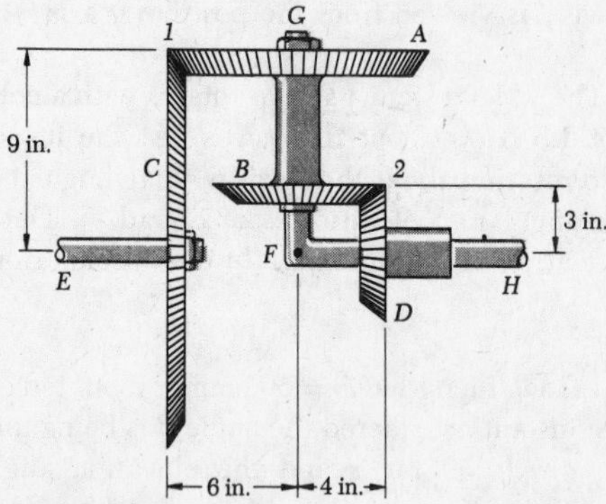

Fig. P15.114

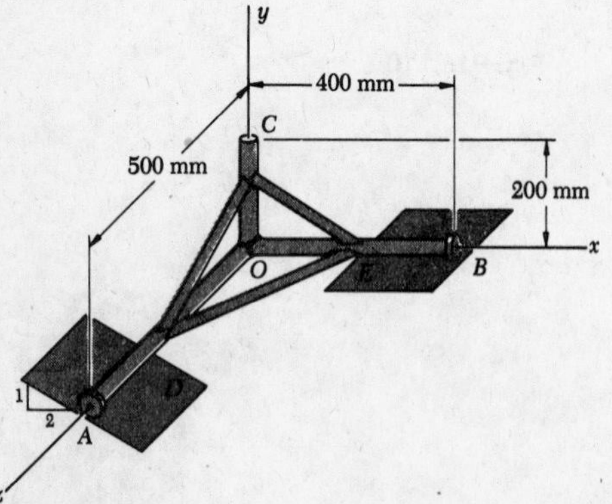

Fig. P15.115

15.115 Three rods are welded together to form the corner assembly shown which is attached to a fixed ball-and-socket joint at O. The end of rod OA moves on the inclined plane D which is perpendicular to the xy plane. The end of rod OB moves on the horizontal plane E which coincides with the zx plane. Knowing that at the instant shown $\mathbf{v}_B = (1.6 \text{ m/s})\mathbf{k}$, determine (*a*) the angular velocity of the assembly, (*b*) the velocity of point C.

15.116 In Prob. 15.114, determine (*a*) the common angular acceleration of gears A and B, (*b*) the acceleration of the tooth of gear B which is in contact with gear D at point 2.

15.117 In Prob. 15.115 the speed of point B is known to be constant. For the position shown, determine (*a*) the angular acceleration of the assembly, (*b*) the acceleration of point C.

15.118 Rod AB, of length 220 mm, is connected by ball-and-socket joints to collars A and B, which slide along the two rods shown. Knowing that collar A moves downward with a constant speed of 63 mm/s, determine the velocity of collar B when $c = 120$ mm.

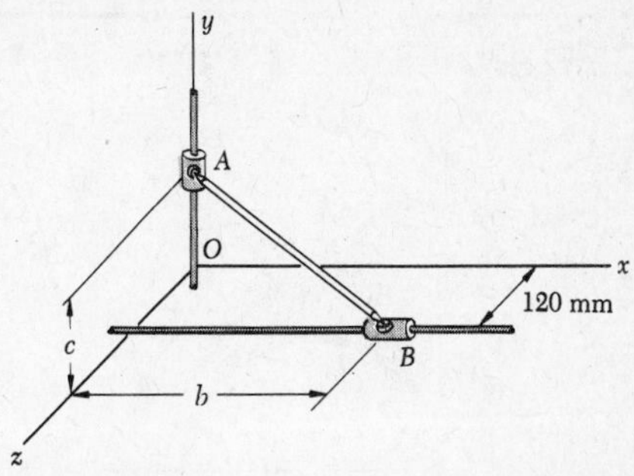

Fig. P15.118

15.119 Solve Prob. 15.118 when $c = 40$ mm.

15.120 Rod BC, of length 21 in., is connected by ball-and-socket joints to the collar C and to the rotating arm AB. Knowing that arm AB rotates in the zx plane at the constant rate $\omega_0 = 38$ rad/s, determine the velocity of collar C.

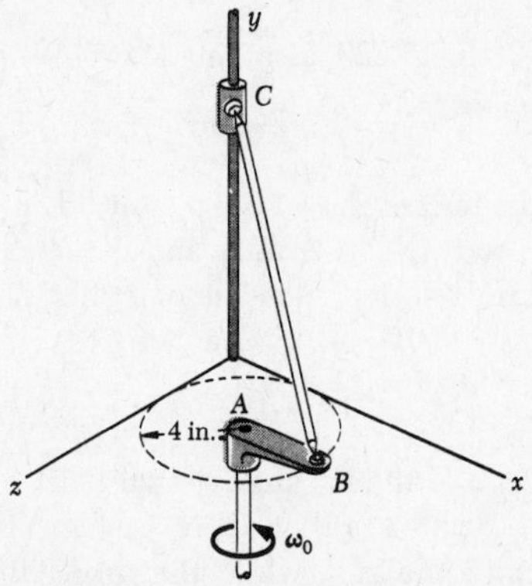

Fig. P15.120

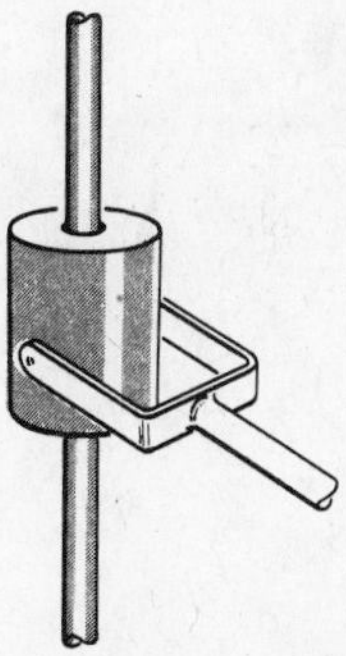

Fig. P15.121 and P15.122

15.121 In Prob. 15.118, the ball-and-socket joint between the rod and collar A is replaced by the clevis connection shown. Determine (*a*) the angular velocity of the rod, (*b*) the velocity of collar B.

15.122 In Prob. 15.120, the ball-and-socket joint between the rod and collar C is replaced by the clevis connection shown. Determine (*a*) the angular velocity of the rod, (*b*) the velocity of collar C.

15.123 In the linkage shown, crank BC rotates in the yz plane while crank ED rotates in a plane parallel to the xy plane. Knowing that in the position shown crank BC has an angular velocity ω_1 of 10 rad/s and no angular acceleration, determine the corresponding angular velocity ω_2 of crank ED.

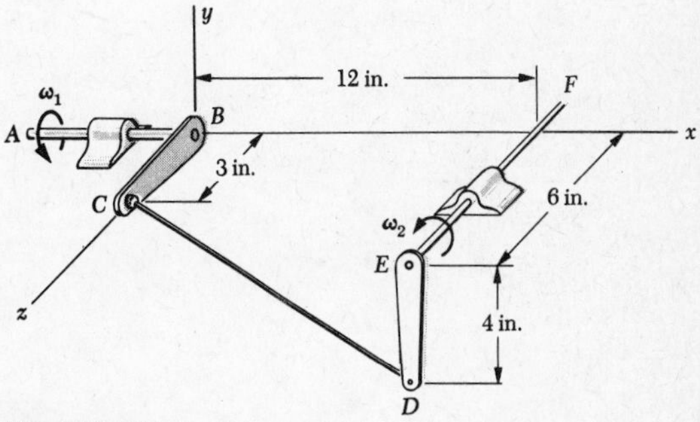

Fig. P15.123

***15.124** In Prob. 15.118, determine the acceleration of collar B when $c = 120$ mm.

***15.125** In Prob. 15.118, determine the acceleration of collar B when $c = 40$ mm.

***15.126** In Prob. 15.120, determine the acceleration of collar C.

***15.127** In Prob. 15.123, determine the angular acceleration of crank ED.

SECTIONS 15.14 and 15.15

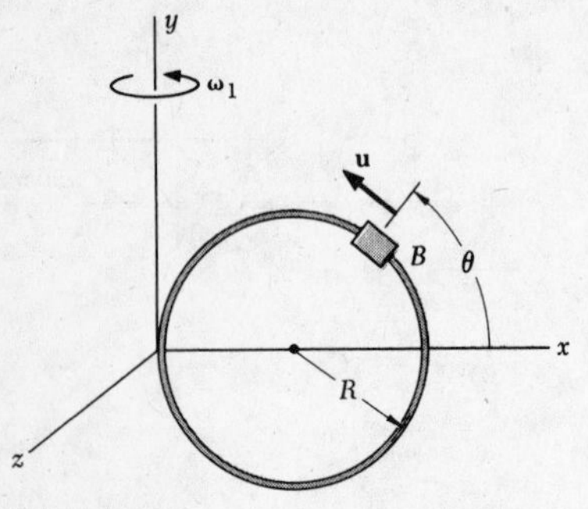

Fig. P15.128

15.128 Collar B is made to move along the ring shown at a speed u relative to the ring. Knowing that the ring rotates about the y axis with a constant angular velocity ω_1, determine the acceleration of B when (a) $\theta = 0$, (b) $\theta = 90°$, (c) $\theta = 180°$.

15.129 The bent rod ABC rotates at a constant rate ω_1. Knowing that the collar D moves downward along the rod at a constant relative speed u, determine for the position shown (a) the velocity of D, (b) the acceleration of D.

15.130 Solve Prob. 15.129, assuming that $\omega_1 = 9$ rad/s, $u = 40$ in./s, and $r = 6$ in.

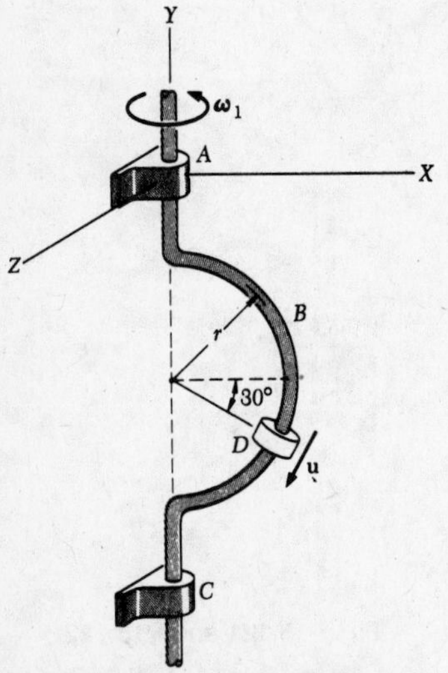

Fig. P15.129 and P15.131

15.131 At the instant shown the magnitude of the angular velocity ω_1 of the bent rod ABC is 9 rad/s and is increasing at the rate of 20 rad/s^2, while the relative speed u of collar D is 40 in./s and is increasing at the rate of 100 in./s^2. Knowing that $r = 6$ in., determine the acceleration of D.

15.132 The cab of the backhoe shown rotates with the constant angular velocity $\omega_1 = (0.4 \text{ rad/s})\mathbf{j}$ about the Y axis. The arm OA is fixed with respect to the cab, while the arm AB rotates about the horizontal axle A at the constant rate $\omega_2 = d\beta/dt = 0.6$ rad/s. Knowing that $\beta = 30°$, determine (a) the angular velocity and angular acceleration of AB, (b) the velocity and acceleration of point B.

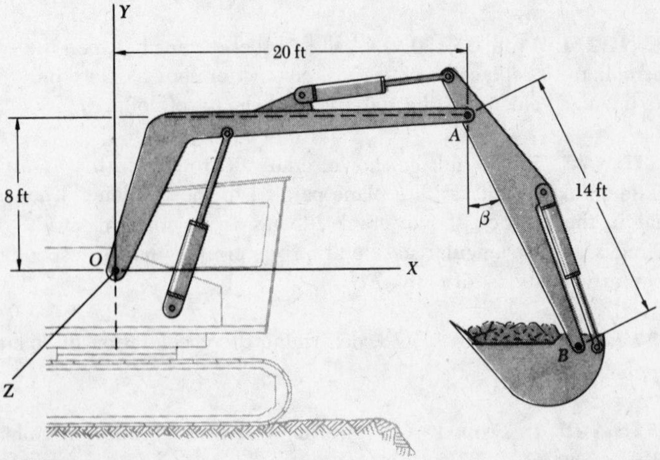

Fig. P15.132

15.133 Solve Prob. 15.132, assuming that $\beta = 30°$ and that arms OA and AB rotate as a rigid body with respect to the cab with a constant angular velocity $(0.6 \text{ rad/s})\mathbf{k}$.

15.134 The elevator B of an ocean liner moves upward with a speed $\dot{R}$ and an acceleration $\ddot{R}$ while the liner is moving to the left with a constant velocity $\mathbf{v}_0$. Knowing that in the position shown the liner rolls at the rate ω_1 and pitches at the rate ω_3 with no angular acceleration, determine (*a*) the velocity of the elevator, (*b*) the acceleration of the elevator. (Note that frame $Oxyz$ is attached to the liner.)

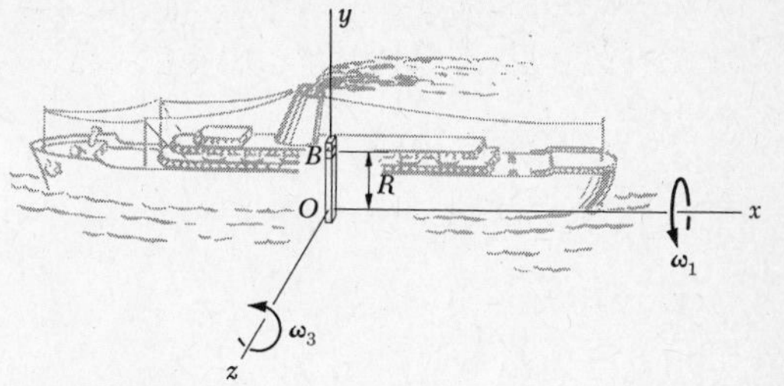

Fig. P15.134

15.135 Solve Sample Prob. 15.14, assuming that the crane has a telescoping boom as shown and that the length of the boom is being increased at the rate $dL/dt = 1.5$ m/s.

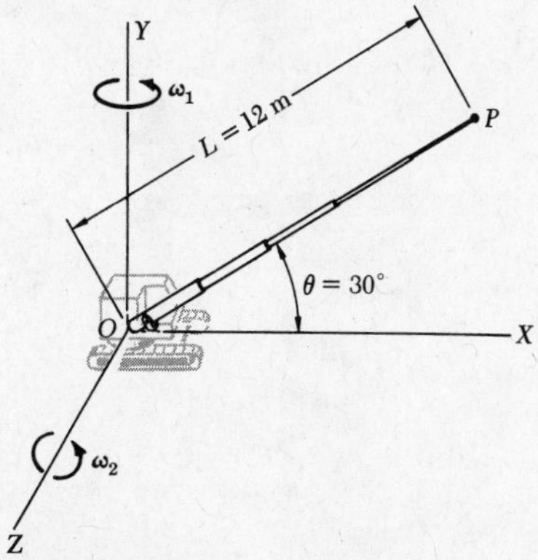

Fig. P15.135

15.136 A disk of radius r rotates at a constant rate ω_2 with respect to the arm CD, which itself rotates at a constant rate ω_1 about the Y axis. Determine (*a*) the angular velocity and angular acceleration of the disk, (*b*) the velocity and acceleration of point B on the rim of the disk.

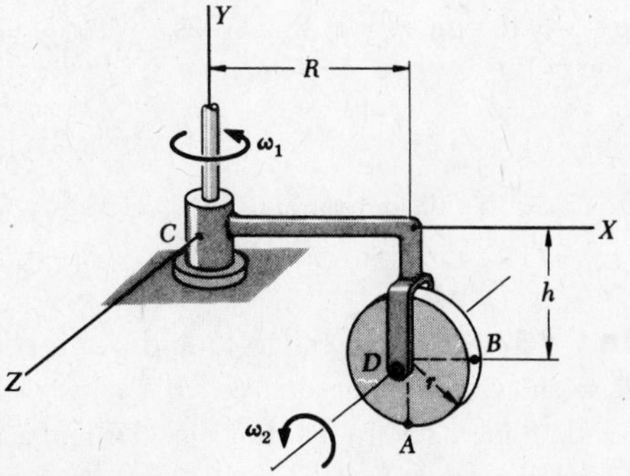

Fig. P15.136

15.137 In Prob. 15.136, determine the velocity and acceleration of point A on the rim of the disk.

15.138 The 40-ft blades of the experimental wind-turbine generator rotate at a constant rate $\omega = 30$ rpm. Knowing that at the instant shown the entire unit is being rotated about the Y axis at a constant rate $\Omega = 0.1$ rad/s, determine (*a*) the angular acceleration of the blades, (*b*) the velocity and acceleration of blade tip B.

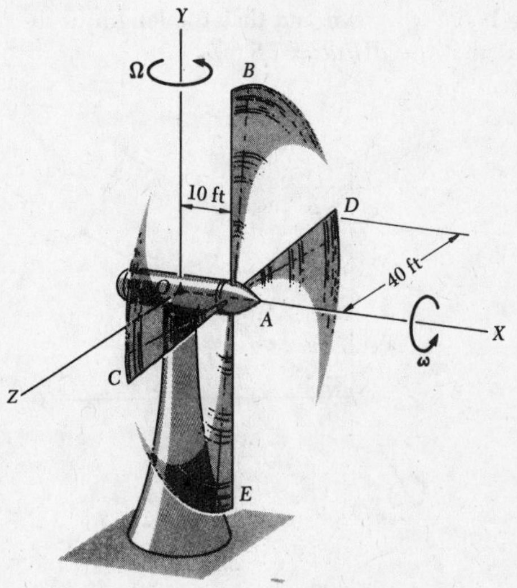

Fig. P15.138

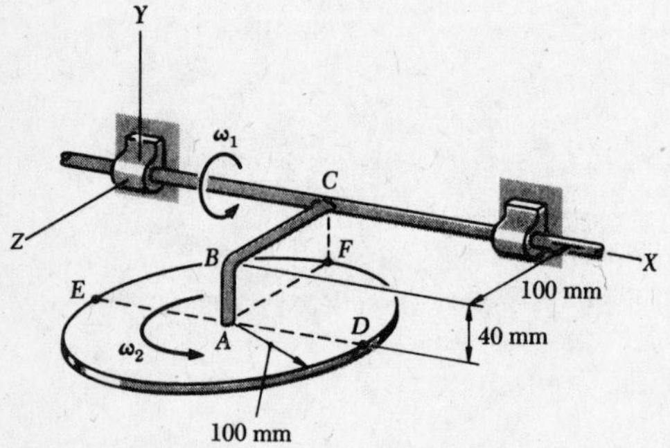

Fig. P15.140

15.139 In Prob. 15.138, determine the velocity and acceleration of (*a*) blade tip C, (*b*) blade tip E.

15.140 A disk of radius 100 mm rotates at a constant rate $\omega_2 = 20$ rad/s with respect to the arm ABC, which itself rotates at a constant rate $\omega_1 = 10$ rad/s about the X axis. Determine (*a*) the angular acceleration of the disk, (*b*) the velocity and acceleration of point D on the rim of the disk.

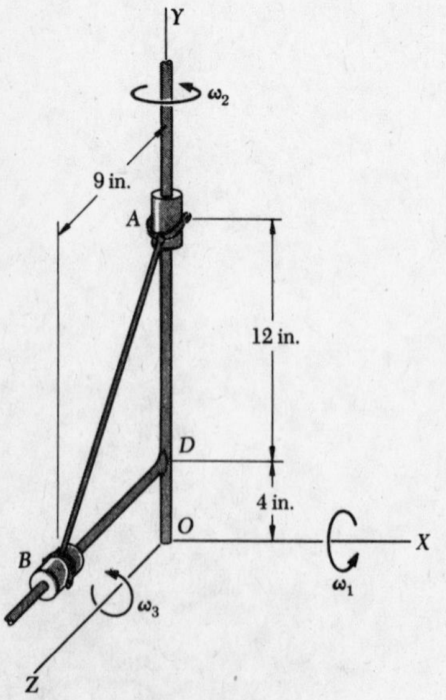

Fig. P15.142 and P15.143

15.141 In Prob. 15.140, determine the acceleration (*a*) of point E, (*b*) of point F.

15.142 and 15.143 Two collars A and B are connected by a 15-in. rod AB as shown. Knowing that collar A moves downward at a constant speed of 18 in./s, determine the velocities and accelerations of collars A and B for the constant rate of rotation indicated.

15.142 $\omega_1 = 10$ rad/s, $\omega_2 = \omega_3 = 0$.

15.143 $\omega_2 = 10$ rad/s, $\omega_1 = \omega_3 = 0$.

CHAPTER 16
PLANE MOTION OF RIGID BODIES: FORCES AND ACCELERATIONS

SECTIONS 16.1 to 16.7

16.1 The open tailgate of a truck is held by hinges at A and a chain as shown. Determine the maximum allowable acceleration of the truck if the tailgate is to remain in the position shown.

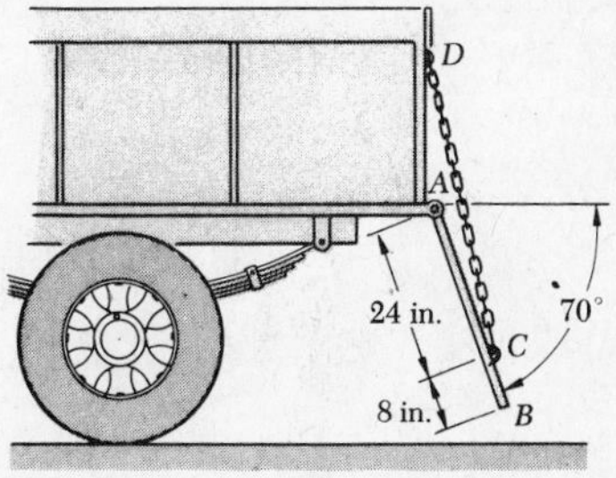

Fig. P16.1

16.2 The motion of a 3-lb semicircular rod is guided by two small wheels which roll freely in a vertical slot. Knowing that the acceleration of the rod is $a = \frac{1}{4}g$ upward, determine (*a*) the magnitude of the force **P**, (*b*) the reactions at A and B.

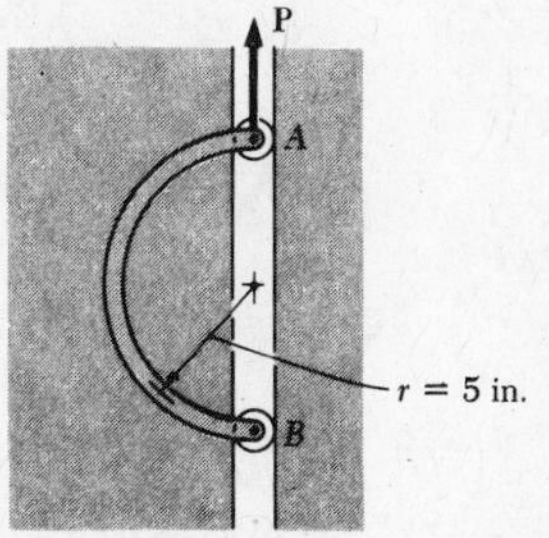

Fig. P16.2

16.3 A 600-kg fork-lift truck carries the 300-kg crate at the height shown. The truck is moving to the left when the brakes are applied causing a deceleration of 3 m/s². Knowing that the coefficient of friction between the crate and the fork lift is 0.5, determine the vertical component of the reaction (*a*) at each of the two wheels A (one wheel on each side of the truck), (*b*) at the single steerable wheel B.

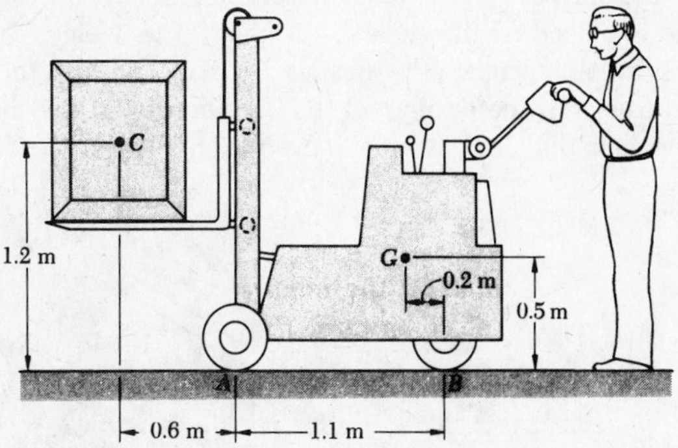

Fig. P16.3

16.4 In Prob. 16.3 determine the maximum deceleration of the truck if the crate is not to slide forward and if the truck is not to tip forward.

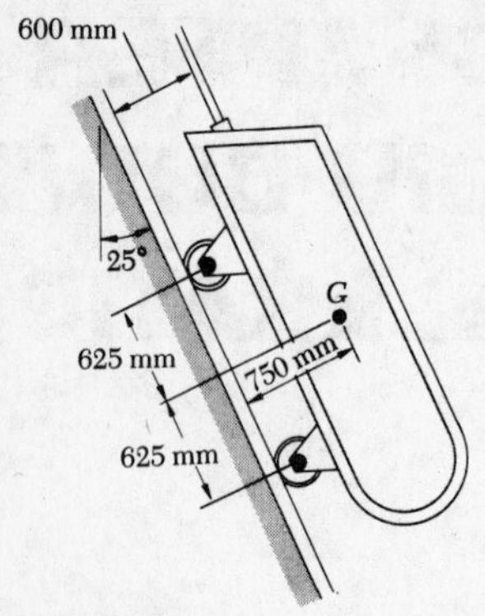

Fig. P16.6

16.5 A man rides a bicycle at a speed of 30 km/h. The distance between axles is 1050 mm, and the mass center of the man and bicycle is located 650 mm behind the front axle and 1000 mm above the ground. If the man applies the brakes on the front wheel only, determine the shortest distance in which he can stop without being thrown over the front wheel.

16.6 The total mass of the loading car and its load is 2500 kg. Neglecting the mass and friction of the wheels, determine (*a*) the minimum tension *T* in the cable for which the upper wheels are lifted from the track, (*b*) the corresponding acceleration of the car.

16.7 A 200-lb rectangular panel is suspended from two skids which may slide with no friction on the inclined track shown. If the panel is released from rest, determine (*a*) the acceleration of the panel, (*b*) the reaction at *B*.

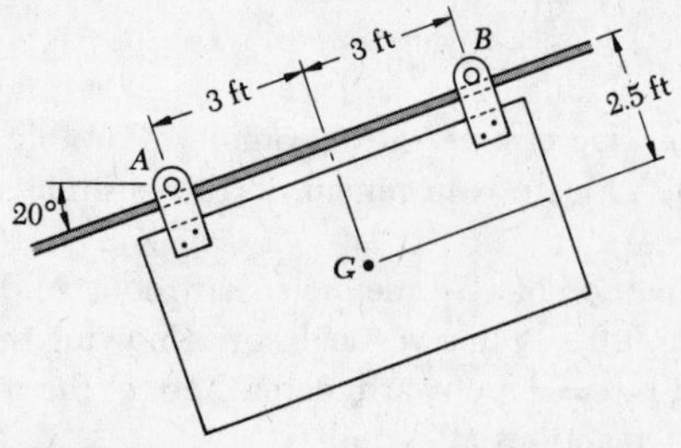

Fig. P16.7

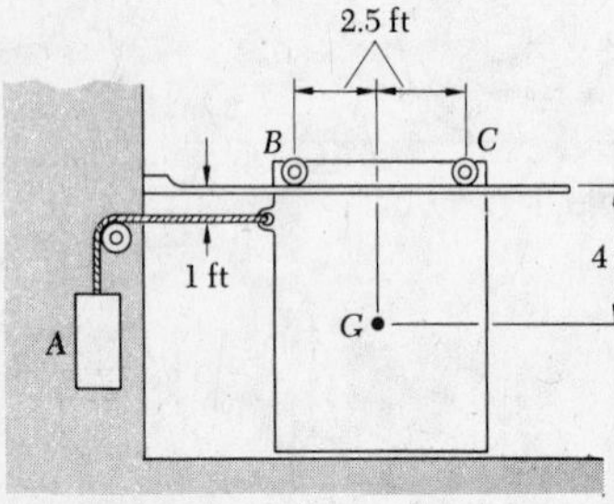

Fig. P16.8

16.8 The 500-lb fire door is supported by wheels *B* and *C* which may roll freely on the horizontal track. The counterweight *A* weighs 100 lb and is connected to the door by the cable shown. If the system is released from rest, determine (*a*) the acceleration of the door, (*b*) the reactions at *B* and *C*.

16.9 The retractable shelf shown is supported by two identical linkage-and-spring systems; only one of the systems is shown. A 40-lb machine is placed on the shelf so that half of its weight is supported by the system shown. If the springs are removed and the system is released from rest, determine (*a*) the acceleration of the machine, (*b*) the tension in link *AB*. Neglect the weight of the shelf and links.

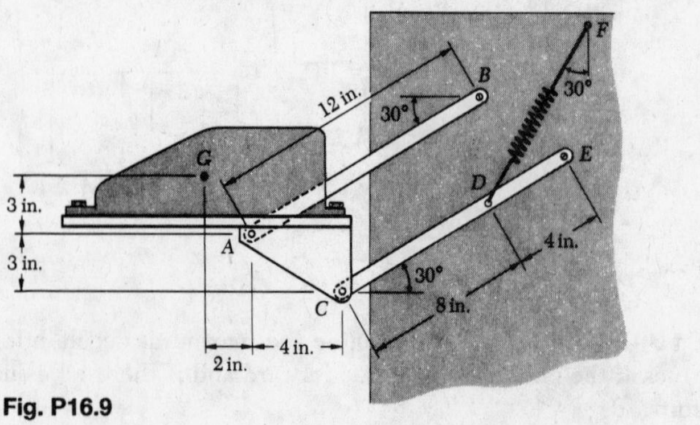

Fig. P16.9

16.10 Two uniform rods AB and BC, each of weight 6 lb, are welded together to form the rigid body ABC which is held in the position shown by the wire AE and the links AD and BE. Determine the force in each link immediately after wire AE has been cut. Neglect the weight of the links.

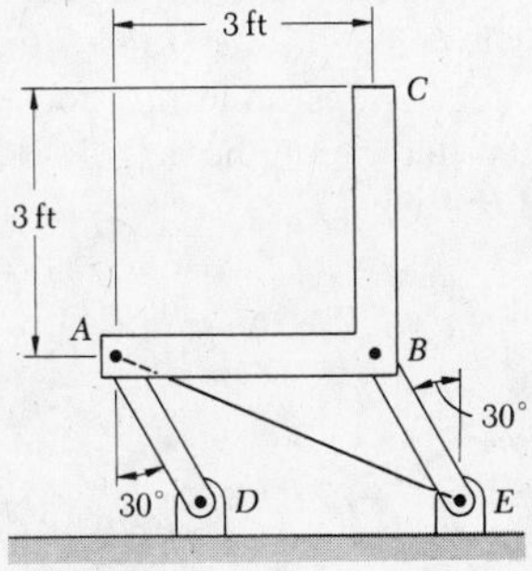

Fig. P16.10

16.11 The cranks AB and CD rotate at a constant speed of 240 rpm. For the position $\phi = 30°$, determine the horizontal components of the forces exerted on the 5-kg uniform connecting rod BC by the pins B and C.

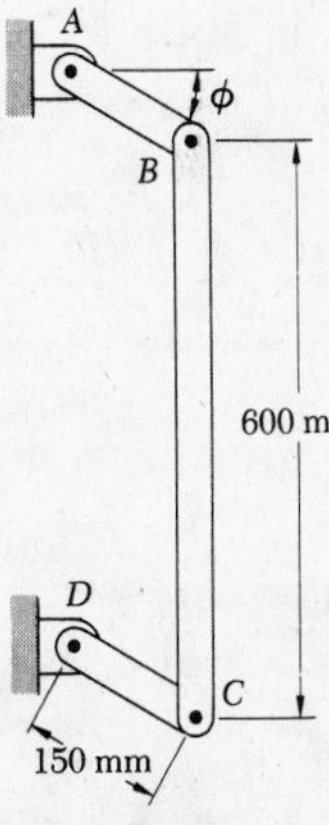

Fig. P16.11

16.12 The control rod AC is guided by two pins which slide freely in parallel curved slots of radius 200 mm. The rod has a mass of 10 kg, and its mass center is located at point G. Knowing that for the position shown the *vertical* component of the velocity of C is 1.25 m/s upward and the *vertical* component of the acceleration of C is 5 m/s² upward, determine the magnitude of the force **P**.

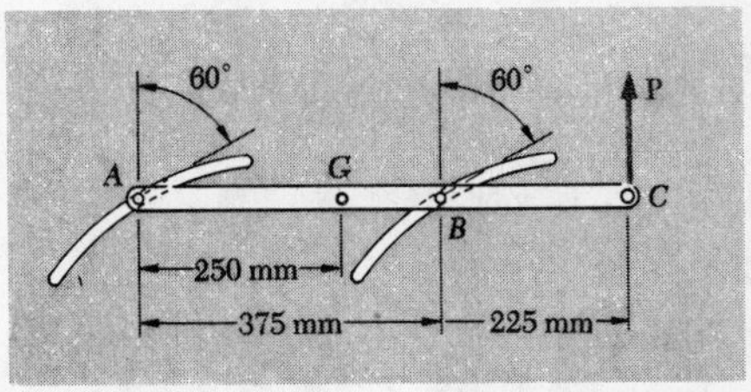

Fig. P16.12

***16.13** Draw the shear and bending-moment diagrams for the rod BC of Prob. 16.11.

***16.14** Draw the shear and bending moment diagrams for the vertical rod BC of Prob. 16.10.

16.15 The 15-in.-radius brake drum is attached to a larger flywheel which is not shown. The total mass moment of inertia of the flywheel and drum is 15 lb · ft · s². Knowing that the initial angular velocity is 150 rpm clockwise, determine the force which must be exerted by the hydraulic cylinder if the system is to stop in 10 revolutions.

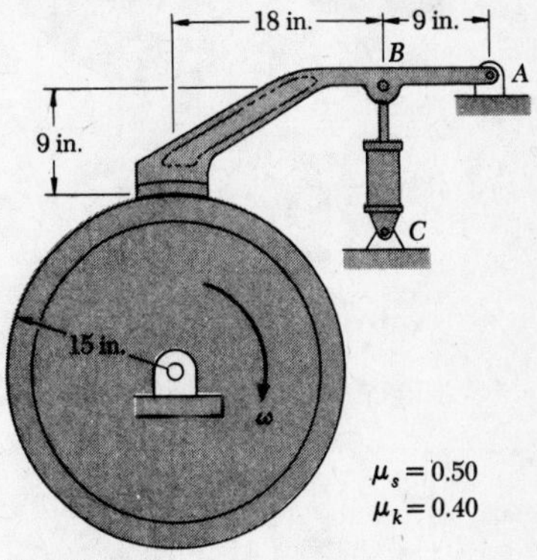

Fig. P16.15

16.16 Solve Prob. 16.15, assuming that the initial angular velocity of the flywheel is 150 rpm counterclockwise.

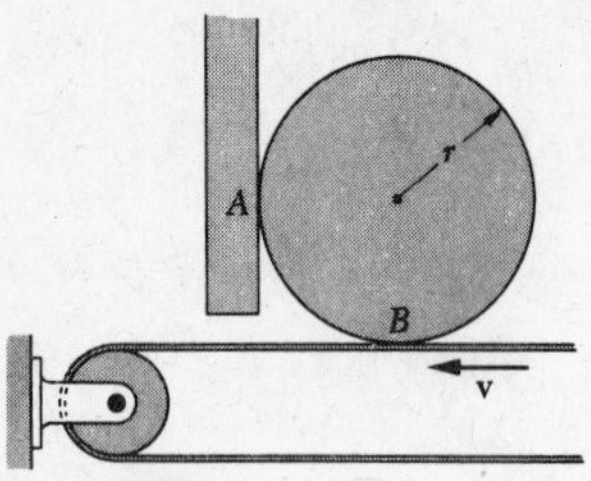

Fig. P16.17

16.17 A cylinder of radius r and mass m is placed with no initial velocity on a belt as shown. Denoting by μ the coefficient of friction at A and at B and assuming that $\mu < 1$, determine the angular acceleration $\boldsymbol{\alpha}$ of the cylinder.

16.18 Shaft A and friction disk B have a combined mass of 15 kg and a combined radius of gyration of 150 mm. Shaft D and friction wheel C rotate with a constant angular velocity of 1000 rpm. Disk B is at rest when it is brought into contact with the rotating wheel. Knowing that disk B accelerates uniformly for 12 s before acquiring its final angular velocity, determine the magnitude of the friction force between the disk and the wheel.

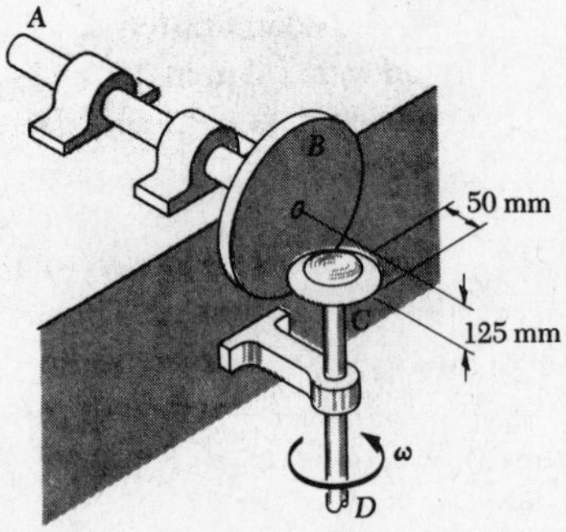

Fig. P16.18

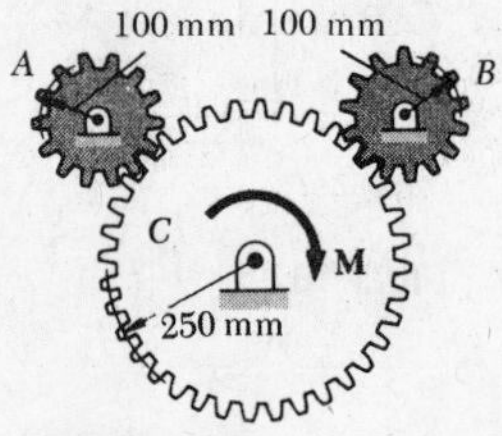

Fig. P16.19

16.19 Each of the gears A and B has a mass of 2 kg and a radius of gyration of 75 mm, while gear C has a mass of 10 kg and a radius of gyration of 225 mm. If a couple **M** of constant magnitude 5 N·m is applied to gear C, determine (a) the angular acceleration of gear A, (b) the time required for the angular velocity of gear A to increase from 150 to 500 rpm.

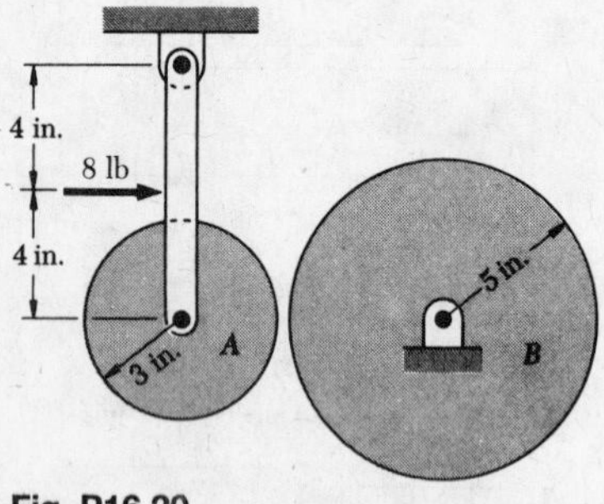

Fig. P16.20

16.20 The two friction disks A and B are brought together by applying the 8-lb force shown. Disk A weighs 6 lb and had an initial angular velocity of 1200 rpm clockwise; disk B weighs 15 lb and was initially at rest. Knowing that $\mu = 0.30$ between the disks and neglecting bearing friction, determine (a) the angular acceleration of each disk, (b) the final angular velocity of each disk.

16.21 Solve Prob. 16.20, assuming that, initially, disk A was at rest and disk B had an angular velocity of 1200 rpm clockwise.

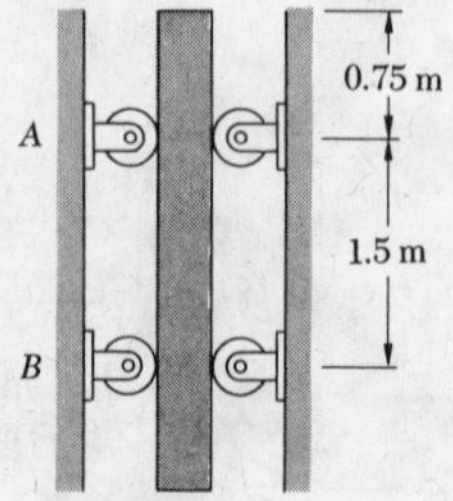

Fig. P16.22

16.22 A 6-kg bar is held between four disks as shown. Each disk has a mass of 3 kg and a diameter of 200 mm. The disks may rotate freely, and the normal reaction exerted by each disk on the bar is sufficient to prevent slipping. If the bar is released from rest, determine (a) its acceleration immediately after release, (b) its velocity after it has dropped 0.75 m

16.23 A 5-m beam of mass 225 kg is lowered from a considerable height by means of two cables unwinding from overhead cranes. As the beam approaches the ground, the crane operators apply brakes to slow the unwinding motion. The deceleration of cable A is 6 m/s^2, while that of cable B is 0.75 m/s^2. Determine the tension in each cable.

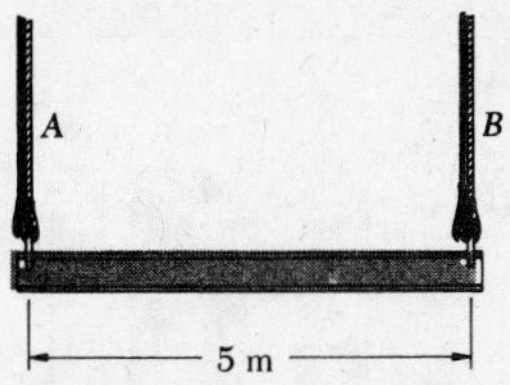

Fig. P16.23 and P16.24

16.24 A 5-m beam of mass 225 kg is lowered from a considerable height by means of two cables unwinding from overhead cranes. As the beam approaches the ground, the crane operators apply brakes to slow the unwinding motion. Determine the acceleration of each cable at that instant, knowing that $T_A = 1600$ N and $T_B = 1450$ N.

16.25 An 18-ft beam weighing 400 lb is lowered from a considerable height by means of two cables unwinding from overhead cranes. As the beam approaches the ground, the crane operators apply brakes to slow the unwinding motion. The deceleration of cable A is 21 ft/s^2, while that of cable B is 3 ft/s^2. Determine the tension in each cable.

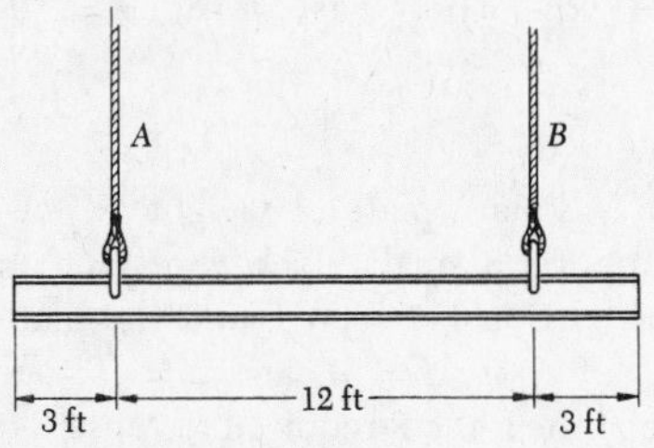

Fig. P16.25 and P16.26

16.26 An 18-ft beam weighing 400 lb is lowered from a considerable height by means of two cables unwinding from overhead cranes. As the beam approaches the ground, the crane operators apply brakes to slow the unwinding motion. Determine the acceleration of each cable at that instant, knowing that $T_A = 325$ lb and $T_B = 275$ lb.

16.27 A turbine disk and shaft have a combined mass of 100 kg and a centroidal radius of gyration of 50 mm. The unit is lifted by two ropes looped around the shaft as shown. Knowing that for each rope $T_A = 270$ N and $T_B = 320$ N, determine (*a*) the angular acceleration of the unit, (*b*) the acceleration of its mass center.

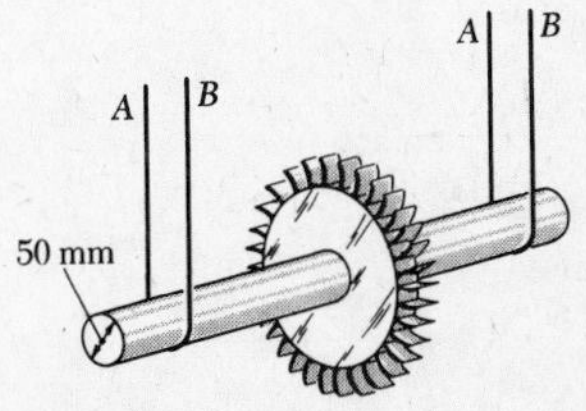

Fig. P16.27

16.28 The 80-lb crate shown rests on four casters which allow it to move without friction in any horizontal direction. A 20-lb horizontal force is applied at the midpoint A of edge CE. Knowing that the force is perpendicular to side $BCDE$, determine the angular acceleration of the crate and the acceleration of point A.

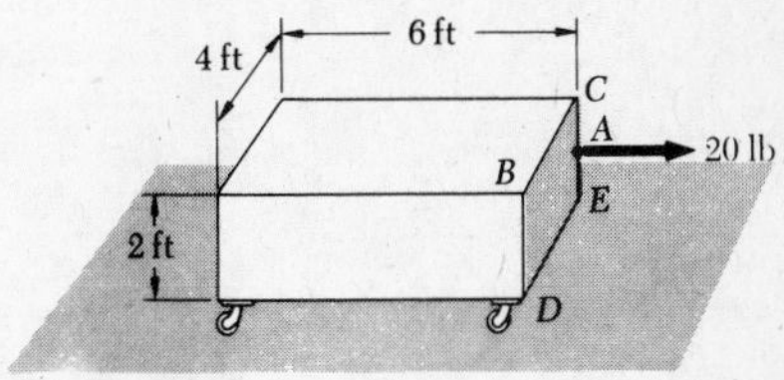

Fig. P16.28

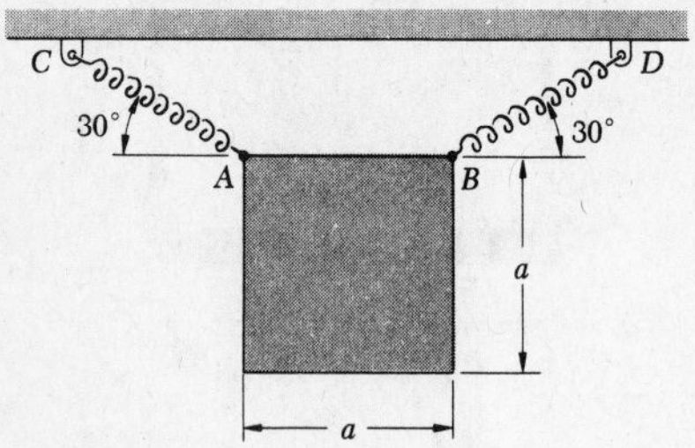

Fig. P16.29

16.29 A uniform square plate of mass m is suspended from two *springs* as shown. If spring BD breaks, determine at that instant (*a*) the angular acceleration of the plate, (*b*) the acceleration of point A, (*c*) the acceleration of point B.

16.30 A disk of mass m and radius r is projected along a rough horizontal surface with a linear velocity $\mathbf{v}_0$ and with $\omega_0 = 0$. The disk will decelerate and then reach a uniform motion. Denoting by μ the coefficient of friction, determine (*a*) the linear and angular acceleration of the disk before it reaches a uniform motion, (*b*) the time required for the motion to become uniform, (*c*) the distance traveled before the motion becomes uniform, (*d*) the final linear and angular velocities of the disk.

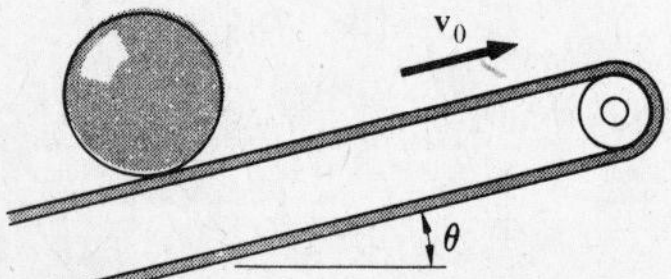

Fig. P16.31

16.31 A sphere of mass m and radius r is dropped with no initial velocity on a belt which moves with a constant velocity $\mathbf{v}_0$. At first the sphere will both slide and roll on the belt. Denoting by μ the coefficient of friction between the sphere and the belt, determine the distance the sphere will move before it starts rolling without sliding.

16.32 Solve Prob. 16.31, assuming $\theta = 10°$, $\mu = 0.30$, and $v_0 = 3$ m/s.

16.33 A heavy square plate of weight W, suspended from four vertical wires, supports a small block E of much smaller weight w. The coefficient of friction between E and the plate is denoted by μ. If the coordinates of E are $x = \frac{1}{4}L$ and $z = \frac{1}{4}L$, derive an expression for the magnitude of the force $\mathbf{P}$ required to cause E to slip with respect to the plate. (*Hint.* Neglect w in all equations containing W.)

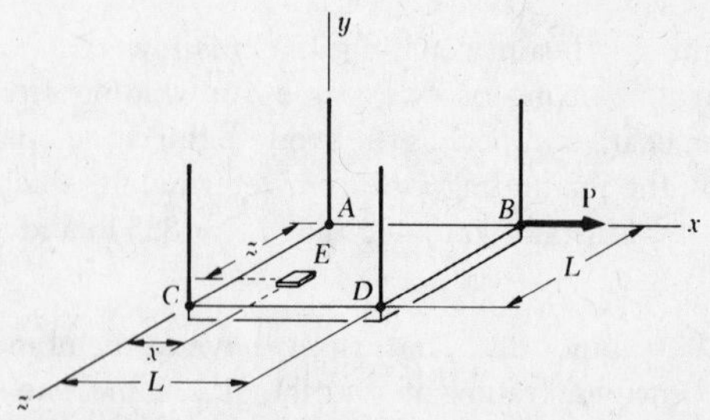

Fig. P16.33 and P16.34

***16.34** A square plate of weight $W = 20$ lb and side $L = 3$ ft is suspended from four wires and supports a block E of much smaller weight w. The coefficient of friction between E and the plate is 0.50. If a force $\mathbf{P}$ of magnitude 10 lb is applied as shown, determine the area of the plate where E should be placed if it is not to slip with respect to the plate. (*Hint.* Neglect w in all equations containing W.)

SECTION 16.8

16.35 A uniform slender rod, of length $L = 900$ mm and mass $m = 4$ kg, is supported as shown. A horizontal force **P** of magnitude 75 N is applied at end B. For $\bar{r} = \frac{1}{4}L = 225$ mm, determine (*a*) the angular acceleration of the rod, (*b*) the components of the reaction at C.

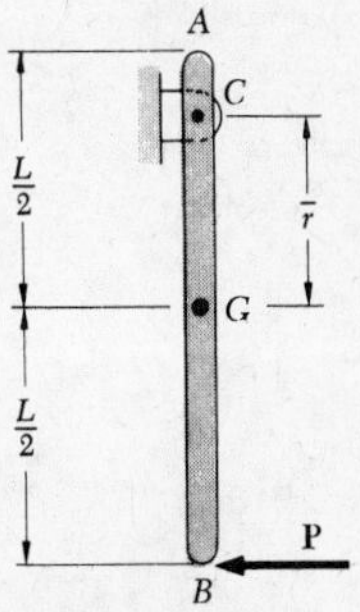

Fig. P16.35

16.36 In Prob. 16.35, determine (*a*) the distance $\bar{r}$ for which the horizontal component of the reaction at C is zero, (*b*) the corresponding angular acceleration of the rod.

16.37 A turbine disk weighing 155 lb rotates at a constant speed of 9600 rpm; the mass center of the disk coincides with the center of rotation O. Determine the reaction at O after a single vane at A, weighing 1.50 oz, becomes loose and is thrown off.

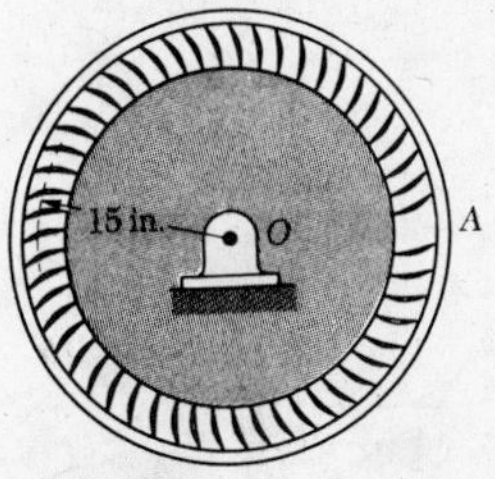

Fig. P16.37

16.38 The rim of a flywheel weighs 4000 lb and has a mean radius of 2 ft. As the flywheel rotates at a constant angular velocity of 360 rpm, radial forces are exerted on the rim by the spokes and internal forces are developed within the rim. Neglecting the weight of the spokes, determine (*a*) the internal forces in the rim, assuming the radial forces exerted by the spokes to be zero, (*b*) the radial force exerted by each spoke, assuming the tangential forces in the rim to be zero.

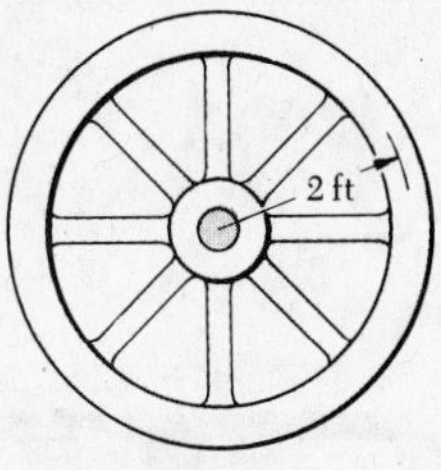

Fig. P16.38

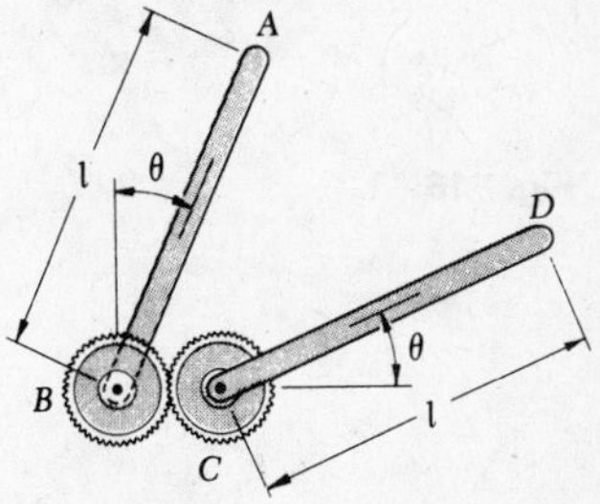

Fig. P16.39

16.39 Two uniform rods, each of mass m, are attached as shown to small gears of negligible mass. If the rods are released from rest in the position shown, determine the angular acceleration of rod AB immediately after release, assuming (*a*) $\theta = 0$, (*b*) $\theta = 30°$.

16.40 A section of pipe, of mass 50 kg and radius 250 mm, rests on two corners as shown. Assuming that μ between the corners and the pipe is sufficient to prevent sliding, determine (*a*) the angular acceleration of the pipe just after corner B is removed, (*b*) the corresponding magnitude of the reaction at A.

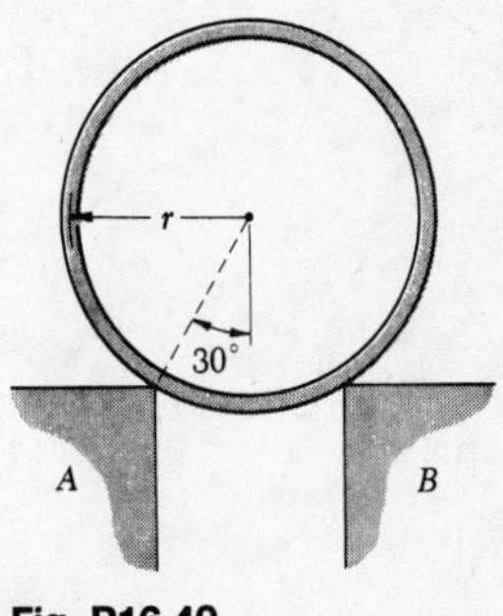

Fig. P16.40

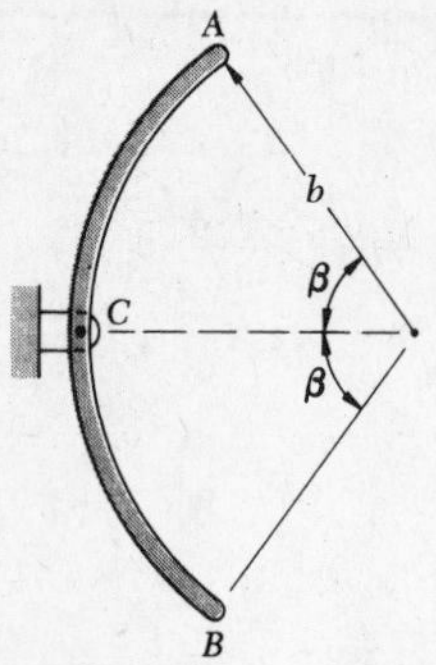

Fig. P16.41

16.41 A uniform rod AB is bent in the shape of an arc of circle. Determine the angular acceleration of the rod immediately after it is released from rest and show that it is independent of β.

16.42 A 12-ft plank, weighing 80 lb, rests on two horizontal pipes AB and CD of a scaffolding. The pipes are 10 ft apart, and the plank overhangs 1 ft at each end. A man weighing 160 lb is standing on the plank when pipe CD suddenly breaks. Determine the initial acceleration of the man, knowing that he was standing (*a*) in the middle of the plank, (*b*) just above pipe CD.

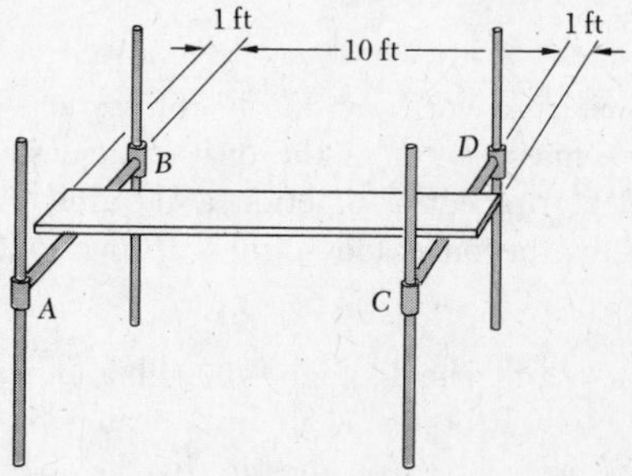

Fig. P16.42

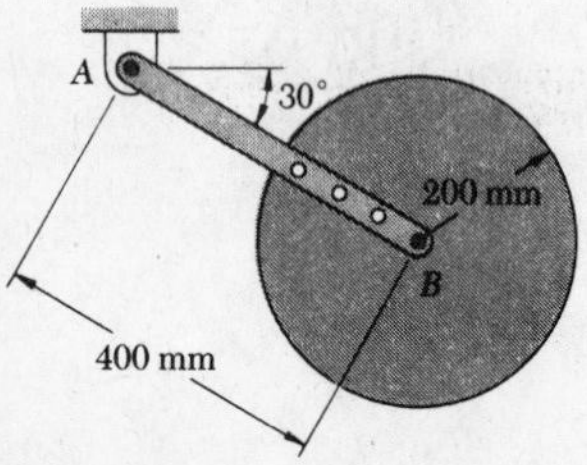

Fig. P16.43

16.43 A 2-kg slender rod is riveted to a 4-kg uniform disk as shown. The assembly swings freely in a vertical plane and, in the position shown, has an angular velocity of 4 rad/s clockwise. Determine (*a*) the angular acceleration of the assembly, (*b*) the components of the reaction at A.

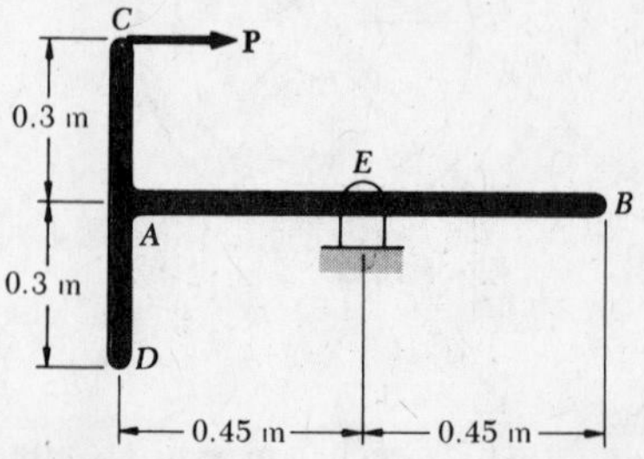

Fig. P16.44

16.44 Two uniform rods, AB of mass 6 kg and CD of mass 4 kg, are welded together to form the T-shaped assembly shown. The assembly rotates in a vertical plane about a horizontal shaft at E. Knowing that at the instant shown the assembly has an angular velocity of 12 rad/s and an angular acceleration of 36 rad/s², both clockwise, determine (*a*) the magnitude of the horizontal force **P**, (*b*) the components of the reaction at E.

16.45 Two uniform rods, AB of weight 6 lb and CD of weight 8 lb, are welded together to form the T-shaped assembly shown. The assembly rotates in a vertical plane about a horizontal shaft at A. Knowing that at the instant shown the assembly has a clockwise angular velocity of 8 rad/s and a counterclockwise angular acceleration of 24 rad/s^2, determine (*a*) the magnitude of the horizontal force **P**, (*b*) the components of the reaction at A.

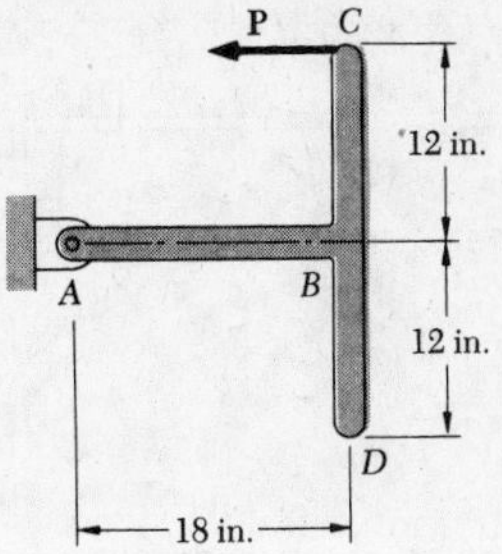

Fig. P16.45

16.46 After being released, the plate of Sample Prob. 16.7 is allowed to swing through 90°. Knowing that at that instant the angular velocity of the plate is 4.82 rad/s, determine (*a*) the angular acceleration of the plate, (*b*) the reaction at A.

16.47 A flywheel is rigidly attached to a shaft of 40-mm radius which may roll along parallel rails as shown. When released from rest, the system rolls a distance of 3 m in 30 s. Determine the centroidal radius of gyration of the system.

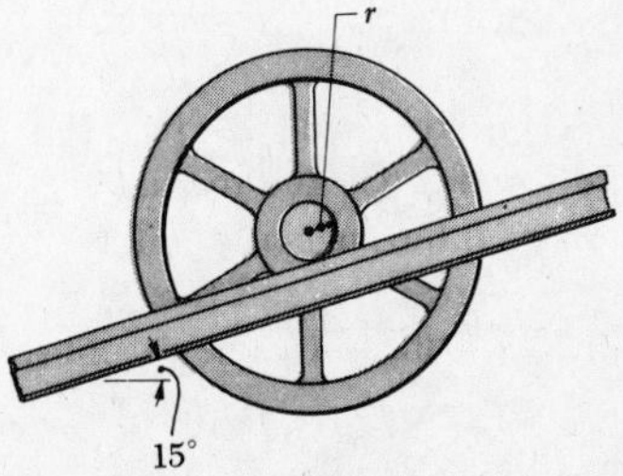

Fig. P16.47 and P16.48

16.48 A flywheel of centroidal radius of gyration $\bar{k} = 600$ mm is rigidly attached to a shaft of radius $r = 30$ mm which may roll along parallel rails. Knowing that the system is released from rest, determine the distance it will roll in 20 s.

16.49 and 16.50 The 12-lb carriage is supported as shown by two uniform disks each of weight 8 lb and radius 3 in. Knowing that the disks roll without sliding, determine the acceleration of the carriage when a force of 4 lb is applied to it.

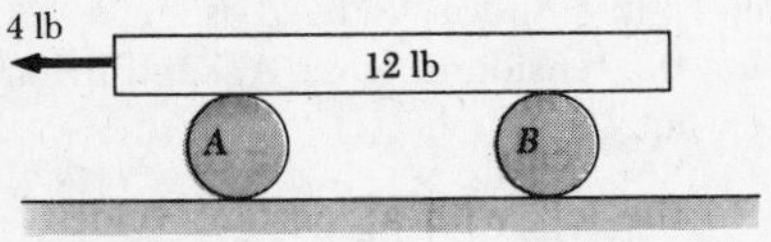

Fig. P16.49

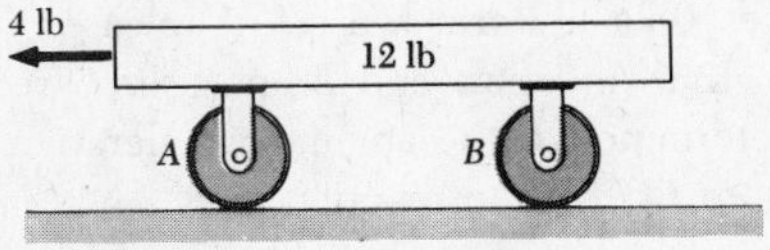

Fig. P16.50

16.51 A half section of pipe of mass m and radius r rests on a rough horizontal surface. A vertical force **P** is applied as shown. Assuming that the section rolls without sliding, derive an expression (*a*) for its angular acceleration, (*b*) for the minimum value of μ compatible with this motion. [*Hint.* Note that $OG = 2r/\pi$ and that, by the parallel-axis theorem, $\bar{I} = mr^2 - m(OG)^2$.]

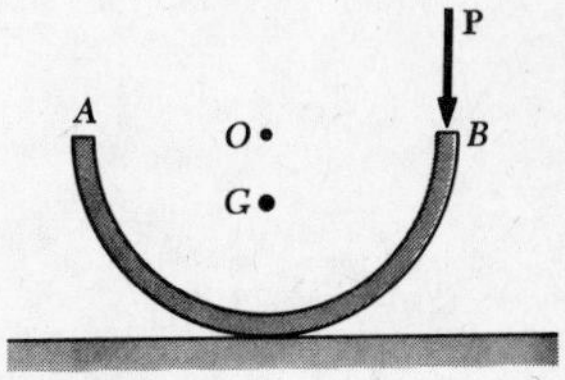

Fig. P16.51

16.52 Solve Prob. 16.51, assuming that the force **P** is applied at B and is directed horizontally to the right.

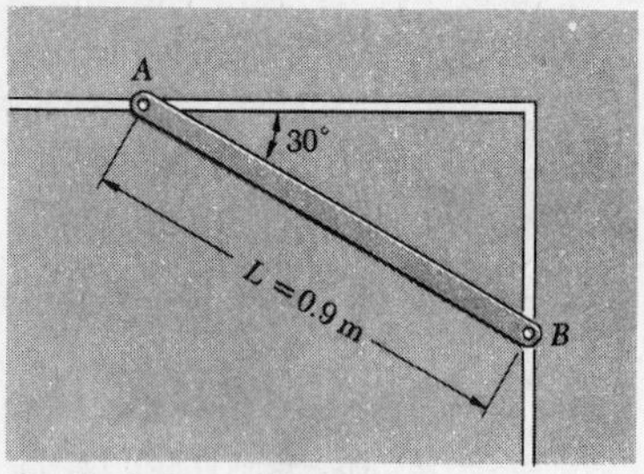

Fig. P16.53 and P16.54

16.53 The motion of a 4-kg slender rod is guided by pins at *A* and *B* which slide freely in the slots shown. A vertical force **P** is applied at *B*, causing the rod to start from rest with a counterclockwise angular acceleration of 12 rad/s^2. Determine (*a*) the magnitude and direction of **P**, (*b*) the reactions at *A* and *B*.

16.54 The motion of a 4-kg slender rod is guided by pins at *A* and *B* which slide freely in the slots shown. A couple **M** is applied to the rod which is initially at rest. Knowing that the initial acceleration of point *B* is 6 m/s^2 upward, determine (*a*) the moment of the couple, (*b*) the reactions at *A* and *B*.

16.55 Rod *AB* weighs 3 lb and is released from rest in the position shown. Assuming that the ends of the rod slide without friction, determine (*a*) the angular acceleration of the rod, (*b*) the reactions at *A* and *B*.

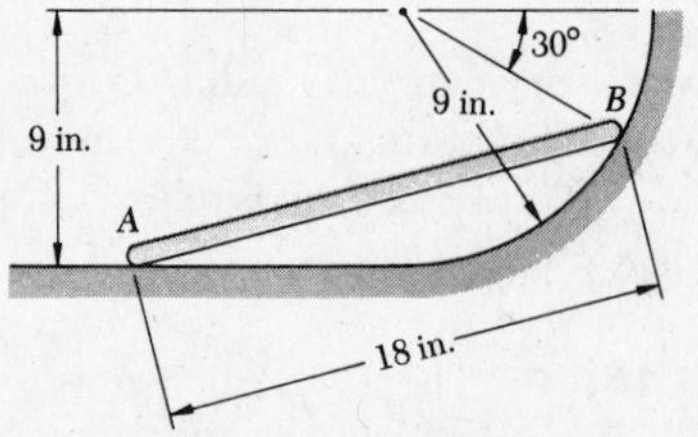

Fig. P16.55

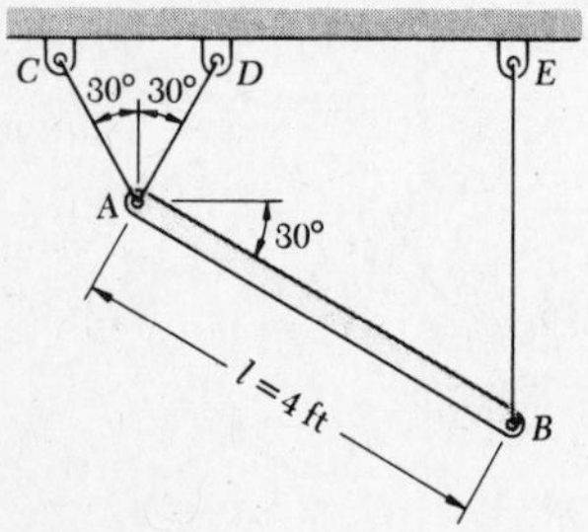

Fig. P16.56

16.56 The 12-lb uniform rod *AB* is held by the three wires shown. Determine the tension in wires *AD* and *BE* immediately after wire *AC* has been cut.

16.57 The rod of Prob. 16.53 is released from rest in the position shown. Determine (*a*) the angular acceleration of the rod, (*b*) the reactions at *A* and *B*.

16.58 The 50-kg uniform rod *AB* is released from rest in the position shown. Knowing that end *A* may slide freely on the frictionless floor, determine (*a*) the angular acceleration of the rod, (*b*) the tension in wire *BC*, (*c*) the reaction at *A*.

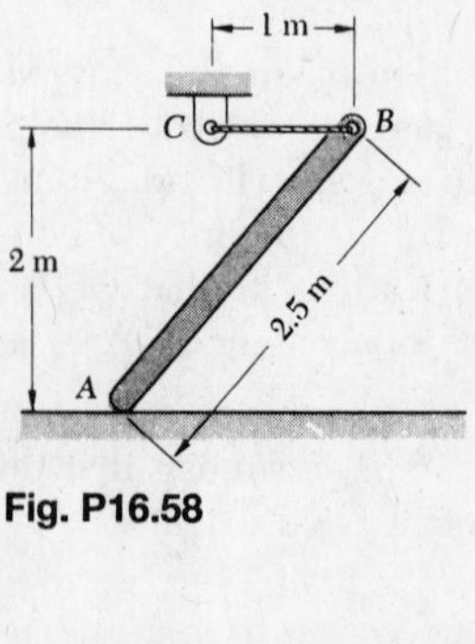

Fig. P16.58

16.59 A 3- by 6-ft crate weighing 200 lb is released from rest in the position shown. Assuming corners A and B slide without friction, determine (*a*) the angular acceleration of the crate, (*b*) the reactions at A and B.

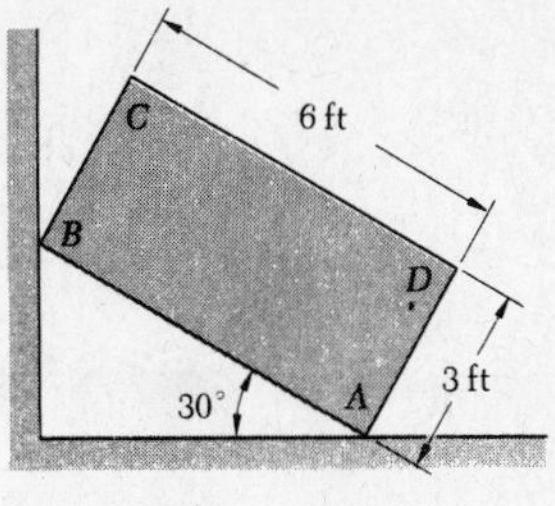

Fig. P16.59

16.60 Two uniform rods AB and BC, each of weight 10 lb, are welded together to form a single 20-lb rigid body which is held by the three wires shown. Determine the tension in wires BD and CF immediately after wire CE has been cut.

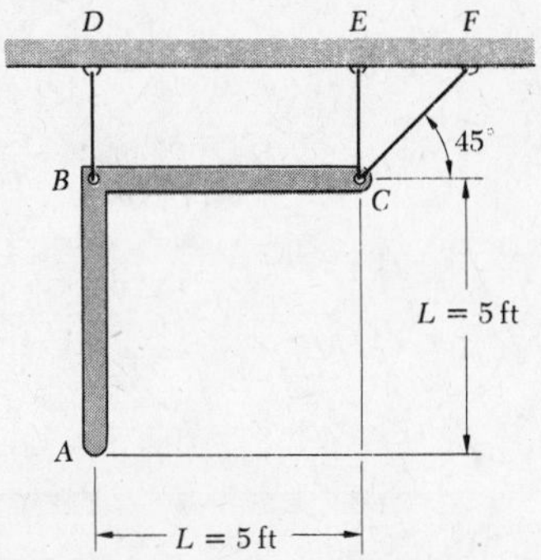

Fig. P16.60

16.61 The 3-kg sliding block is connected to the rotating disk by the uniform rod AB of mass 2 kg. Knowing that the disk has a constant angular velocity of 360 rpm, determine the forces exerted on the connecting rod at A and B when $\beta = 0$. Assume the motion to take place in a horizontal plane.

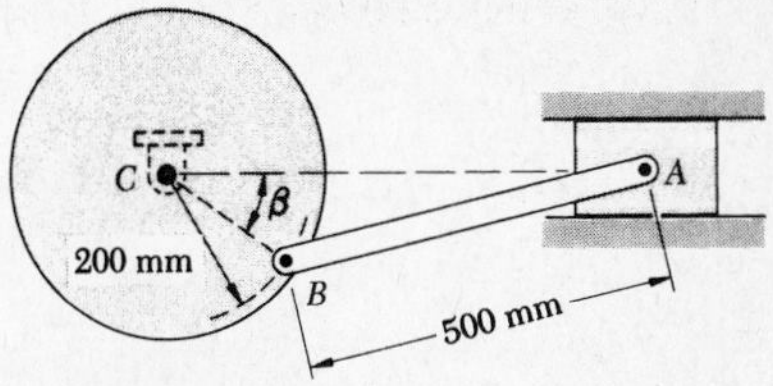

Fig. P16.61

16.62 Solve Prob. 16.61 when $\beta = 180°$.

16.63 End A of the 100-lb beam AB moves along the frictionless floor, while end B is supported by a 4-ft cable. Knowing that at the instant shown end A is moving to the left with a constant velocity of 8 ft/s, determine (*a*) the magnitude of the force **P**, (*b*) the corresponding tension in the cable.

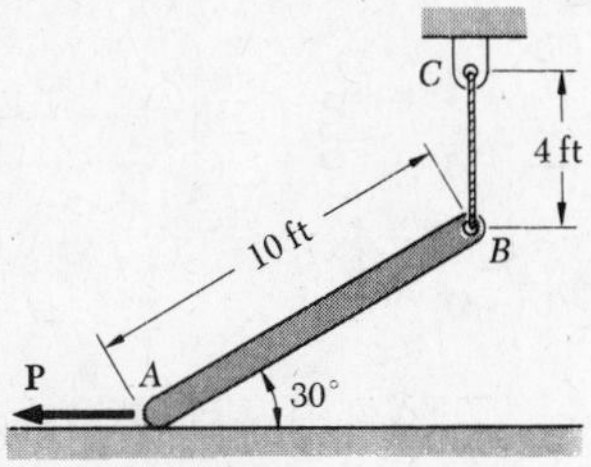

Fig. P16.63

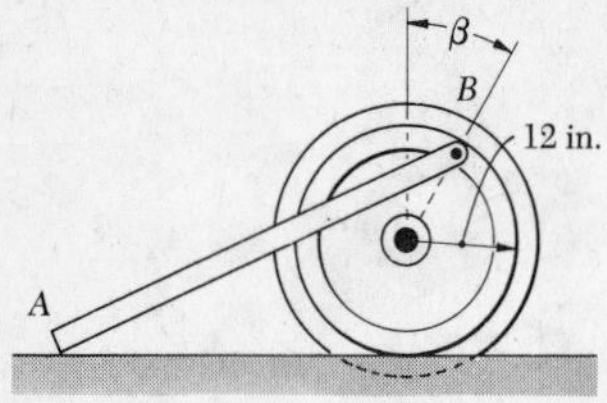

Fig. P16.64

16.64 The flanged wheel shown rolls to the right with a constant velocity of 4 ft/sec. The rod AB is 4 ft long and weighs 10 lb. Knowing that point A slides without friction on the horizontal surface, determine the reaction at A (*a*) when $\beta = 0$, (*b*) when $\beta = 180°$.

16.65 Each of the bars shown is 600 mm long and has a mass of 4 kg. A horizontal and variable force **P** is applied at C, causing point C to move to the left with a constant speed of 10 m/s. Determine the force **P** for the position shown.

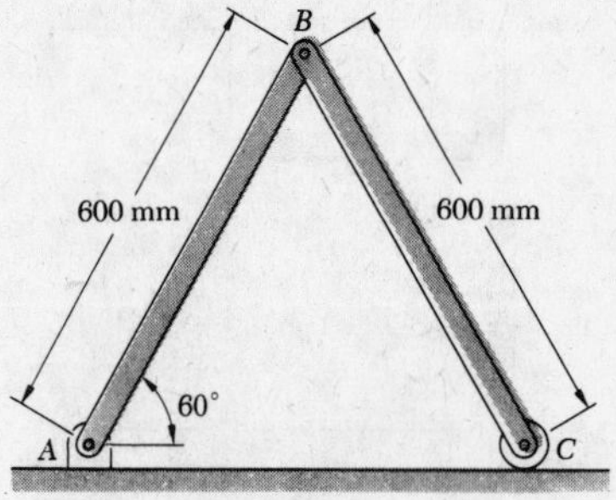

Fig. P16.65 and P16.66

16.66 The two bars AB and BC are released from rest in the position shown. Each bar is 600 mm long and has a mass of 4 kg. Determine (*a*) the angular acceleration of each bar, (*b*) the reactions at A and C.

16.67 and 16.68 Gear C weighs 6 lb and has a centroidal radius of gyration of 3 in. The uniform bar AB weighs 5 lb and gear D is stationary. If the system is released from rest in the position shown, determine (*a*) the angular acceleration of gear C, (*b*) the acceleration of point B.

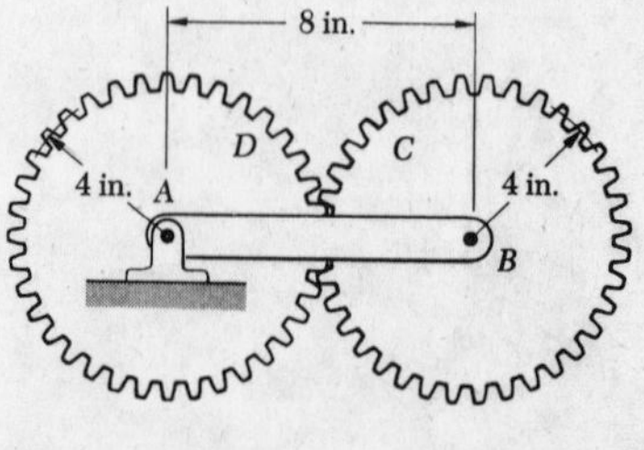

Fig. P16.67

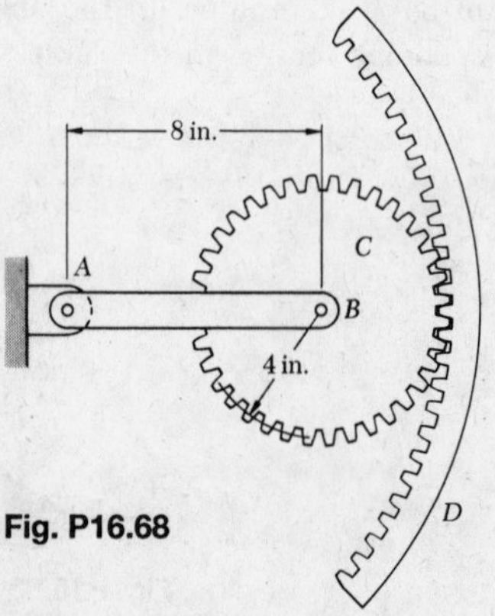

Fig. P16.68

***16.69** The slender bar AB is of length L and mass m. It is held in equilibrium by two counterweights, each of mass $\frac{1}{2}m$. If the wire at B is cut, determine at that instant the acceleration of (*a*) point A, (*b*) point B.

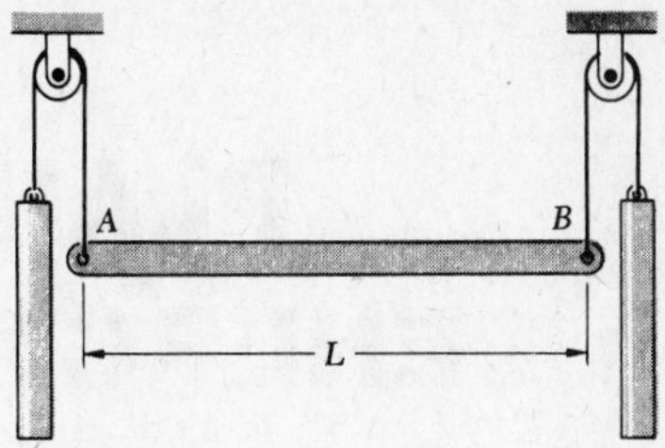

Fig. P16.69

***16.70** Two uniform slender rods AB and CD, each of mass m, are connected by two wires as shown. If the wire BD is cut, determine at that instant the acceleration (*a*) of point D, (*b*) of point B.

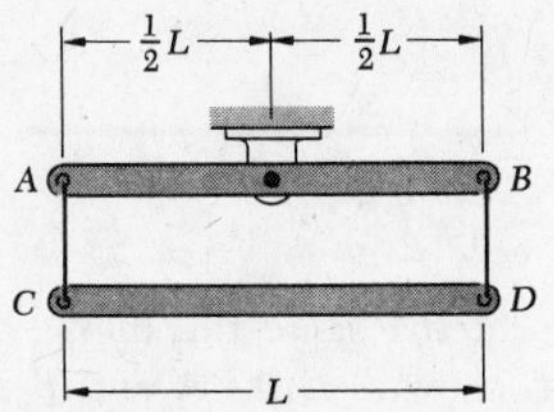

Fig. P16.70

***16.71** Each of the bars AB and BC is of length $L = 2$ ft and weight 5 lb. A couple $\mathbf{M}_0$ of moment 6 lb · ft is applied to bar BC. Determine the angular acceleration of each bar.

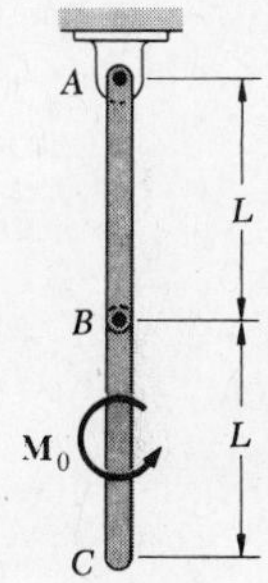

Fig. P16.71

***16.72** Two uniform rods, each of weight W, are released from rest in the position shown. Determine the reaction at A immediately after release.

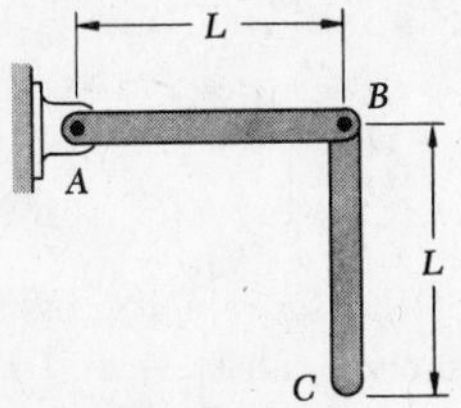

Fig. P16.72

***16.73** Draw the shear and bending-moment diagrams for the bar of Prob. 16.69 immediately after the wire at B has been cut.

***16.74** Solve Prob. 16.72, assuming that the connection at B is welded and that the rods move as a single rigid body. Also, determine the corresponding bending moment at B.

***16.75** In Prob. 16.71 the pin at B is severely rusted and the bars rotate as a single rigid body. Determine the bending moment which occurs at B.

CHAPTER 17
PLANE MOTION OF RIGID BODIES: ENERGY AND MOMENTUM METHODS

SECTIONS 17.1 to 17.7

17.1 Three disks of the same thickness and same material are attached to a shaft as shown. Disks *A* and *B* each have a radius *r*; disk *C* has a radius *nr*. A couple **M** of constant moment is applied when the system is at rest and is removed after the system has executed one revolution. Determine the radius of disk *C* which results in the largest final speed of a point on the rim of disk *C*.

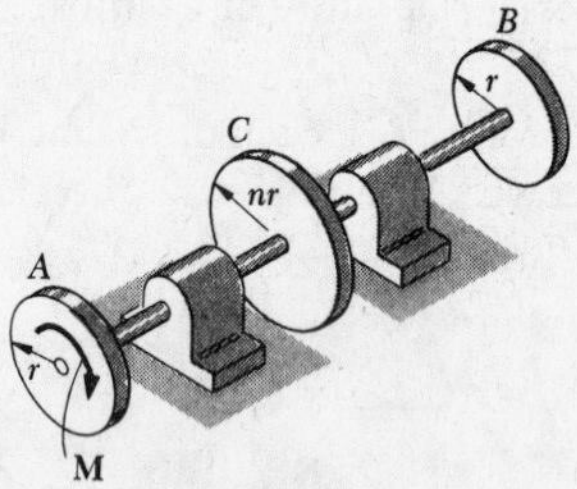

Fig. P17.1 and P17.2

17.2 Three disks of the same thickness and same material are attached to a shaft as shown. Disks *A* and *B* weigh 15 lb each and are of radius $r = 10$ in. A couple **M** of moment 30 lb · ft is applied to disk *A* when the system is at rest. Determine the radius *nr* of disk *C* if the angular velocity of the system is to be 600 rpm after 10 revolutions.

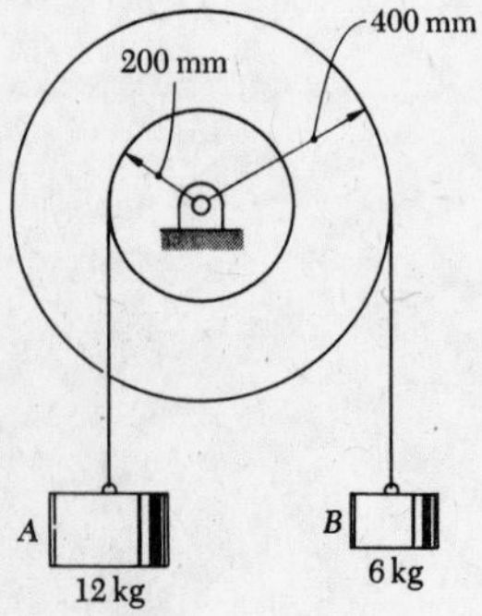

Fig. P17.3

17.3 Two cylinders are attached by cords to an 18-kg double pulley which has a radius of gyration of 300 mm. When the system is at rest and in equilibrium, a 3-kg collar is added to the 12-kg cylinder. Neglecting friction, determine the velocity of each cylinder after the pulley has completed one revolution.

17.4 Solve Prob. 17.3, assuming that the 3-kg collar is added to the 6-kg cylinder.

17.5 The 200-mm-radius brake drum is attached to a larger flywheel which is not shown. The total mass moment of inertia of the flywheel and drum is 8 kg·m^2. Knowing that the initial angular velocity is 120 rpm clockwise, determine the force **P** which must be applied if the system is to come to rest in 8 revolutions.

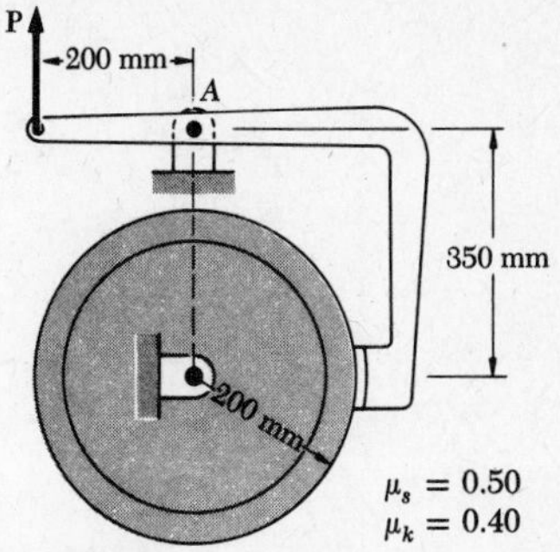

Fig. P17.5

17.6 Solve Prob. 17.5, assuming that the initial angular velocity of the flywheel is 120 rpm counterclockwise.

17.7 The block C of mass m acquires a speed v after being released from rest and falling a distance h. Assuming that no slipping occurs, derive a formula for the speed v in terms of h, m, the radii r_A and r_B, and the centroidal mass moments of inertia $\bar{I}_A$ and $\bar{I}_B$ of the two disks.

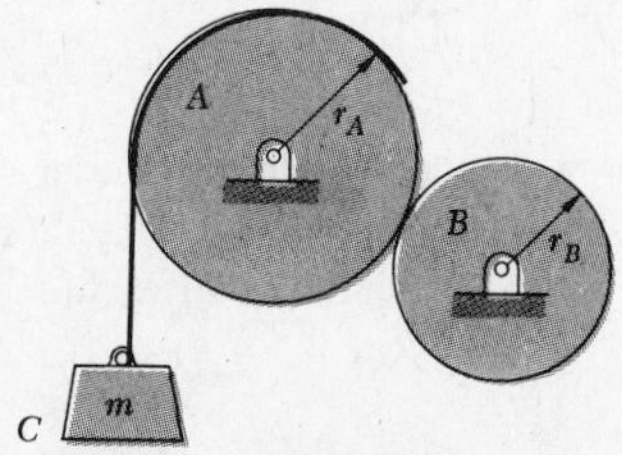

Fig. P17.7 and P17.8

17.8 Disk A weighs 8 lb, and its radius is $r_A = 10$ in.; disk B weighs 3 lb and its radius is $r_B = 6$ in. Knowing that the system is released from rest, determine the velocity of the 2-lb block C after it has fallen 5 ft.

17.9 Disk A is attached to a motor and is made to rotate with a constant angular velocity of 300 rpm clockwise. Disk B has a mass of 5 kg and is at rest when it is placed in contact with disk A. Knowing that $\mu = 0.25$ and neglecting bearing friction, determine the number of revolutions executed by disk B before it reaches a constant angular velocity.

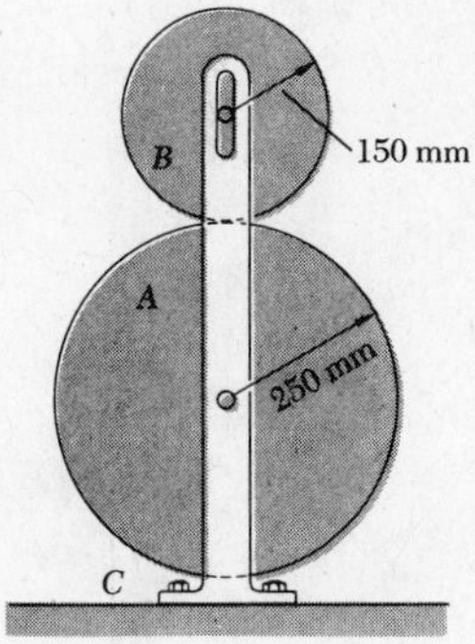

Fig. P17.9

17.10 A uniform rod of length l is pivoted about a point C located at a distance b from its center G. The rod is released from rest in a horizontal position. Determine (*a*) the distance b so that the angular velocity of the rod as it passes through a vertical position is maximum, (*b*) the value of the maximum angular velocity.

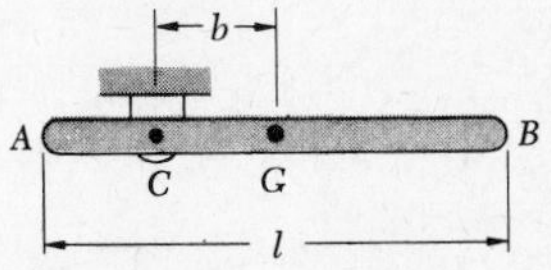

Fig. P17.10

17.11 A 25-lb disk is attached to a 10-lb rod AB as shown. The system is released from rest in the position shown. Determine the velocity of point B as it passes through its lowest position, (*a*) assuming that the disk is welded to the rod, (*b*) assuming that the disk is attached to the rod by a smooth pin at B.

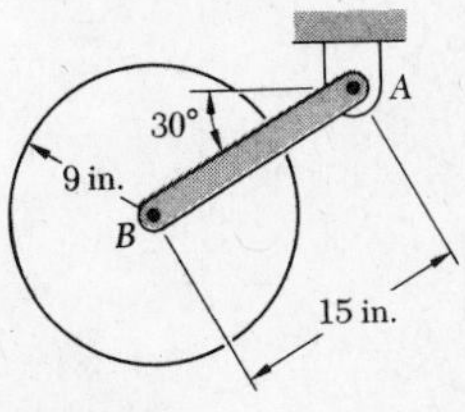

Fig. P17.11

17.12 A 6- by 8-in. rectangular plate is suspended by two pins at A and B. The pin at B is removed and the plate swings about point A. Determine (*a*) the angular velocity of the plate after it has rotated through 90°, (*b*) the maximum angular velocity attained by the plate as it swings freely.

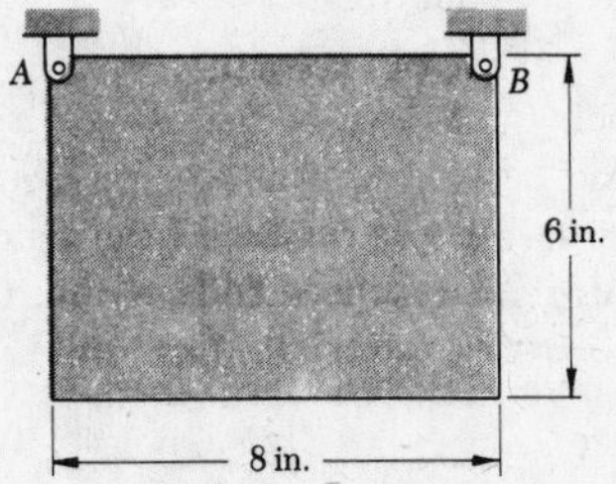

Fig. P17.12

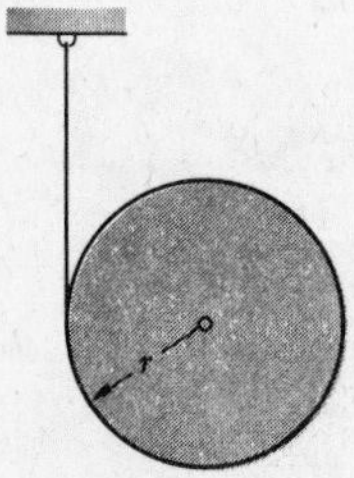

Fig. P17.13

17.13 A cord is wrapped around a cylinder of radius r and mass m as shown. If the cylinder is released from rest, determine the velocity of the center of the cylinder after it has moved through a distance h.

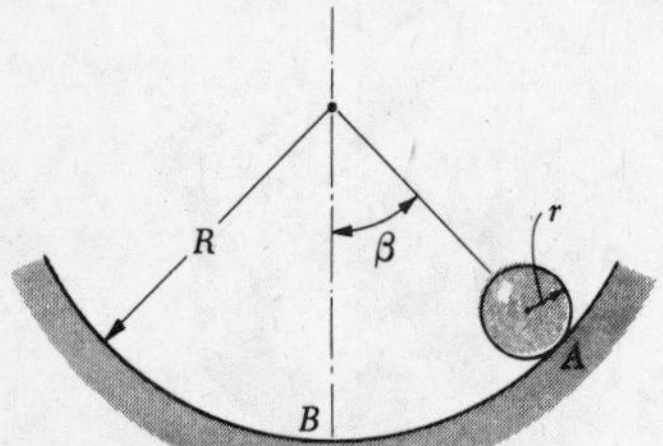

Fig. P17.14

17.14 A sphere of mass m and radius r rolls without slipping inside a curved surface of radius R. Knowing that the sphere is released from rest in the position shown, derive an expression (*a*) for the linear velocity of the sphere as it passes through B, (*b*) for the magnitude of the vertical reaction at that instant.

17.15 and 17.16 The 12-lb carriage is supported as shown by two uniform disks, each of weight 8 lb and radius 3 in. Knowing that the system is initially at rest, determine the velocity of the carriage after it has moved 3 ft. Assume that the disks roll without sliding.

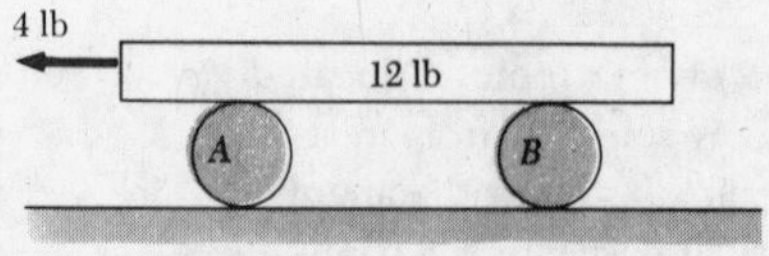

Fig. P17.15

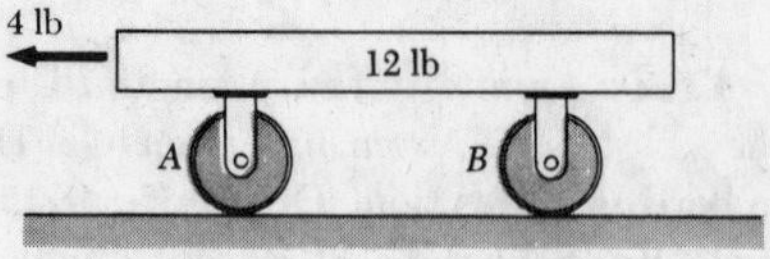

Fig. P17.16

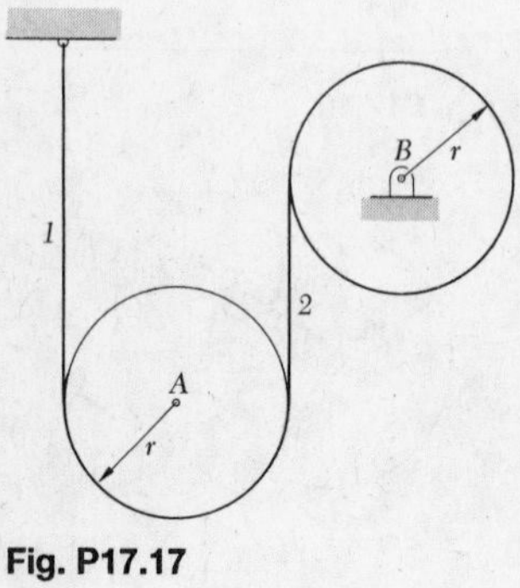

Fig. P17.17

17.17 Two identical disks, each of weight W and radius r, are connected by a cord as shown. If the system is released from rest, determine (*a*) the velocity of the center of disk A after it has moved through a distance s, (*b*) the tension in portion 2 of the cord.

17.18 The mass center G of a 1.5-kg wheel of radius $R = 150$ mm is located at a distance $r = 50$ mm from its geometric center C. The centroidal radius of gyration of the wheel is $\bar{k} = 75$ mm. As the wheel rolls without sliding, its angular velocity is observed to vary. Knowing that in position *1* the angular velocity is 10 rad/s, determine the angular velocity of the wheel (*a*) in position *2*, (*b*) in position *3*.

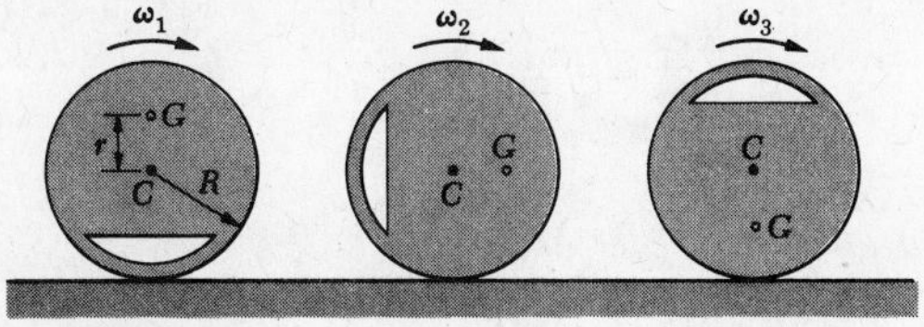

Fig. P17.18

17.19 The motion of the 240-mm rod AB is guided by pins at A and B which slide freely in the slots shown. If the rod is released from rest in position *1*, determine the velocity of A and B when the rod is (*a*) in position *2*, (*b*) in position *3*.

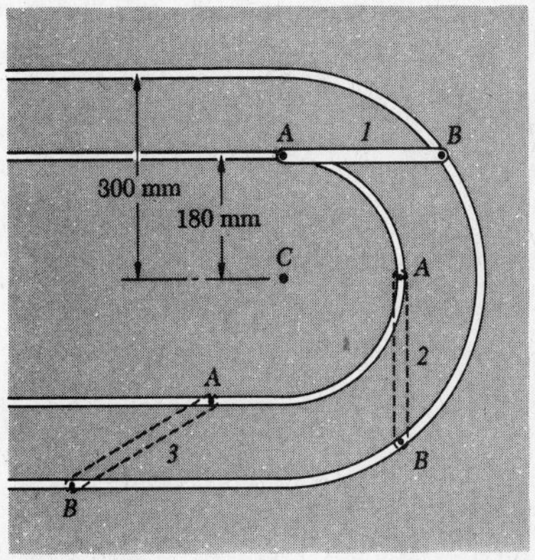

Fig. P17.19

17.20 The motion of a 0.6-m slender rod is guided by pins at A and B which slide freely in the slots shown. Knowing that the rod is released from rest when $\theta = 0$ and that end A is given a slight push to the right, determine (*a*) the angle θ for which the speed of end A is maximum, (*b*) the corresponding maximum speed of A.

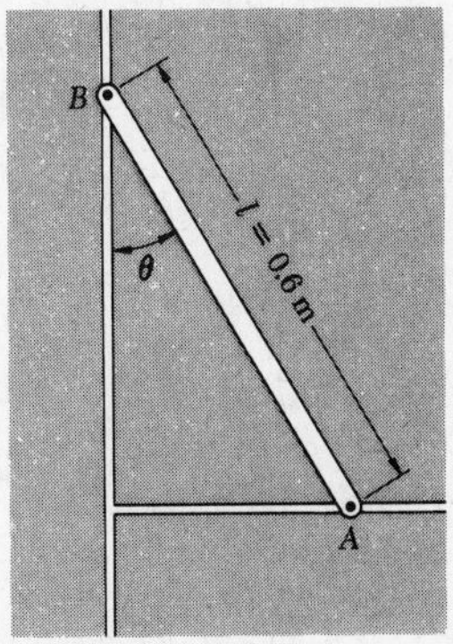

Fig. P17.20

17.21 The motion of a slender rod of length L is guided by pins at A and B which slide freely in the slots shown. If end B is moved slightly to the right and then released, determine (*a*) the angular velocity of the rod and the velocity of the center of the rod as end A reaches point C, (*b*) the position in which the rod will come again to rest.

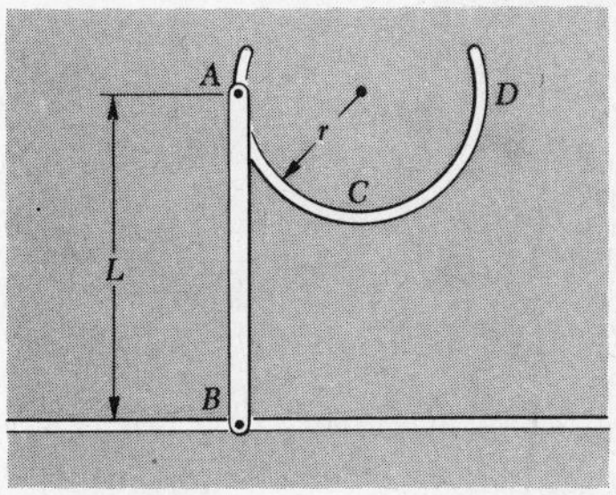

Fig. P17.21

17.22 Solve Prob. 17.21, assuming that end B is moved slightly to the left and then released.

17.23 The uniform rods AB and BC are of mass 4.5 and 1.5 kg respectively. If the system is released from rest in the position shown, determine the angular velocity of rod BC as it passes through a vertical position.

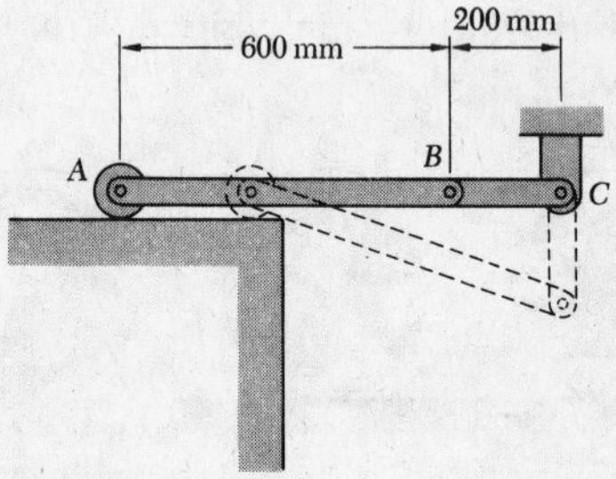

Fig. P17.23

17.24 In Prob. 17.23, determine the angular velocity of rod BC after it has rotated 45°.

17.25 Each of the two rods shown is of length $L = 3$ ft and weighs $W = 5$ lb. Point D is constrained to move along a vertical line. If the system is released from rest when rod AB is horizontal, determine the velocity of points B and D as rod BD passes through a horizontal position.

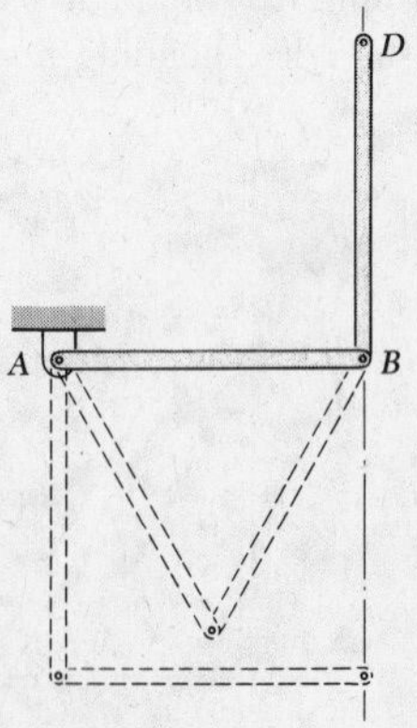

Fig. P17.25

17.26 In Prob. 17.25, determine the velocity of points B and D when points A and D are at the same elevation.

17.27 Knowing that the maximum allowable couple which can be applied to a shaft is 12 kN · m, determine the maximum power which can be transmitted by the shaft (*a*) at 100 rpm, (*b*) at 1000 rpm.

17.28 Determine the moment of the couple which must be exerted by a motor to develop $\frac{1}{4}$ hp at a speed of (*a*) 3600 rpm, (*b*) 720 rpm.

SECTIONS 17.8 to 17.10

17.29 A turbine-generator unit is shut off when its rotor is rotating at 3600 rpm; it is observed that the rotor coasts to rest in 7.10 min. Knowing that the 1850-kg rotor has a radius of gyration of 234 mm, determine the average magnitude of the couple due to bearing friction.

17.30 The rotor of a steam turbine has an angular velocity of 5400 rpm when the steam supply is suddenly cut off. The rotor weighs 200 lb and has a radius of gyration of 6 in. If kinetic friction produces a couple of magnitude 24 lb · in., determine the time required for the rotor to coast to rest.

17.31 A cylinder of radius r and mass m is placed in a corner with an initial counterclockwise angular velocity ω_0. Denoting by μ the coefficient of friction at A and B, derive an expression for the time required for the cylinder to come to rest.

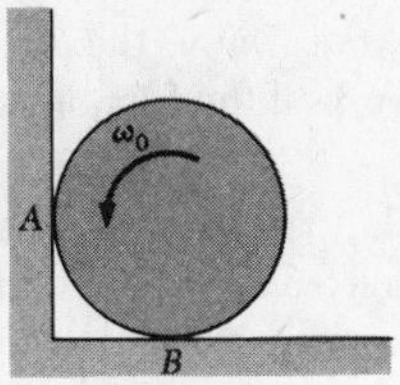

Fig. P17.31

17.32 Solve Prob. 17.31, assuming that the surface at A is frictionless.

17.33 Disks A and B are of mass 5 and 1.8 kg, respectively. The disks are initially at rest and the coefficient of friction between them is 0.20. A couple **M** of magnitude 4 N · m is applied to disk A for 1.50 s and then removed. Determine (a) whether slipping occurs between the disks, (b) the final angular velocity of each disk.

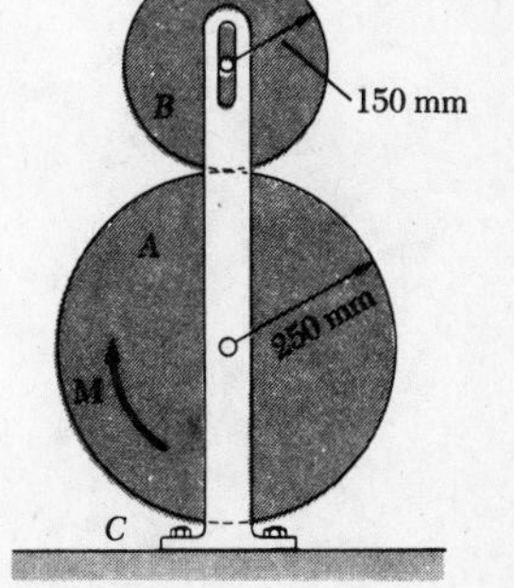

Fig. P17.33

17.34 In Prob. 17.33, determine (a) the largest couple **M** for which no slipping occurs, (b) the corresponding final angular velocity of each disk.

17.35 Two disks A and B are connected by a belt as shown. Each disk weighs 30 lb and has a radius of 1.5 ft. The shaft of disk B rests in a slotted bearing and is held by a spring which exerts a constant force of 15 lb. If a 20-lb·ft couple is applied to disk A, determine (a) the time required for the disks to attain a speed of 600 rpm, (b) the tension in both portions of the belt, (c) the minimum coefficient of friction if no slipping is to occur.

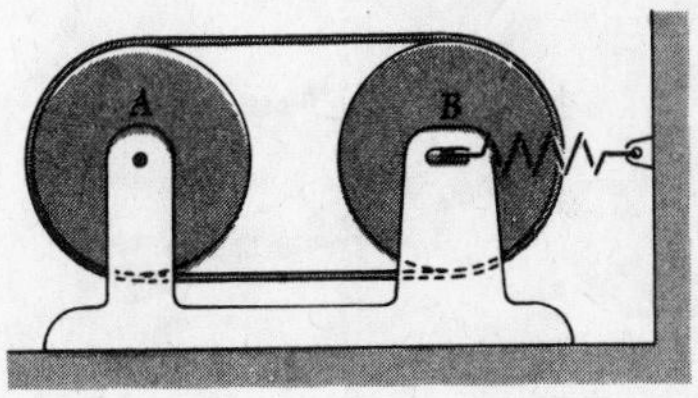

Fig. P17.35

17.36 Solve Prob. 17.35, assuming that disk A weighs 10 lb and disk B weighs 50 lb.

17.37 A section of thin-walled pipe of radius r is released from rest at time $t = 0$. Assuming that the pipe rolls without slipping, determine (*a*) the velocity of the center at time t, (*b*) the coefficient of friction required to prevent slipping.

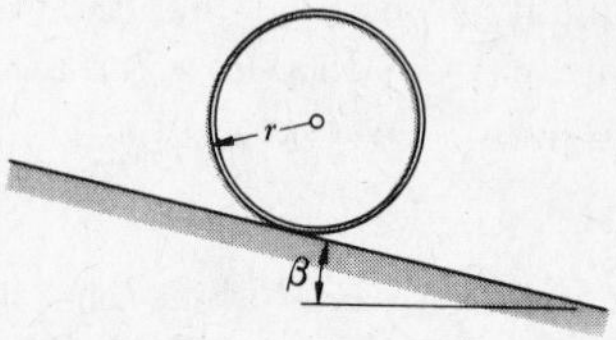

Fig. P17.37

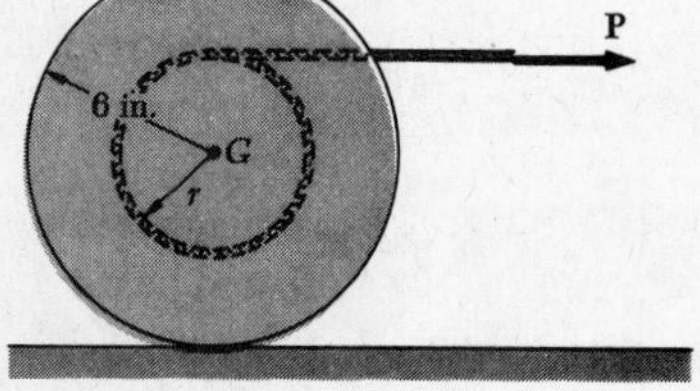

Fig. P17.38

17.38 Two disks, each of weight 12 lb and radius 6 in., which roll without slipping, are connected by a drum of radius r and of negligible weight. A rope is wrapped around the drum and is pulled horizontally with a force **P** of magnitude 8 lb. Knowing that $r = 3$ in. and that the disks are initially at rest, determine (*a*) the velocity of the center G after 3 s, (*b*) the friction force required to prevent slipping.

17.39 In Prob. 17.38, determine the required value of r and the corresponding velocity after 3 s if the friction force is to be zero.

17.40 Four rectangular panels, each of length b and height $\frac{1}{2}b$, are attached with hinges to a circular plate of diameter $\sqrt{2}b$ and held by a wire loop in the position shown. The plate and the panels are made of the same material and have the same thickness. The entire assembly is rotating with an angular velocity ω_0 when the wire breaks. Determine the angular velocity of the assembly after the panels have come to rest in a horizontal position.

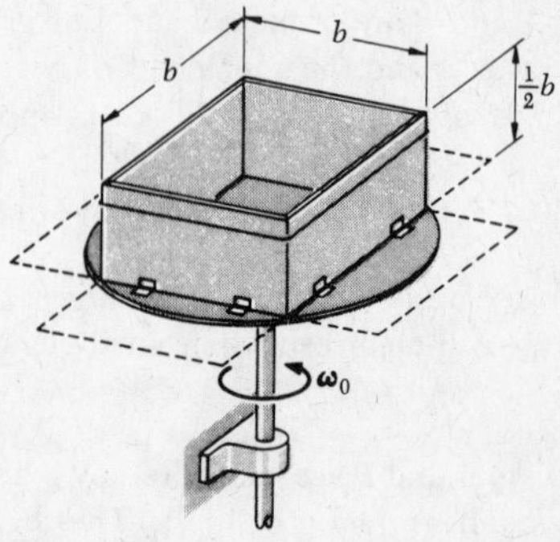

Fig. P17.40

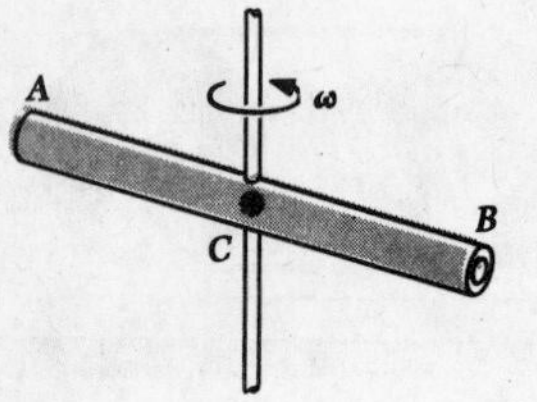

Fig. P17.41

17.41 A small 250-g ball may slide in a slender tube of length 1 m and of mass 1 kg which rotates freely about a vertical axis passing through its center C. If the angular velocity of the tube is 10 rad/s as the ball passes through C, determine the angular velocity of the tube (*a*) just before the ball leaves the tube, (*b*) just after the ball has left the tube.

17.42 A 4-kg tube CD may slide freely on rod AB, which in turn may rotate freely in a horizontal plane. At the instant shown, the assembly is rotating with an angular velocity of magnitude $\omega = 8$ rad/s and the tube is moving toward A with a speed of 1.5 m/s relative to the rod. Knowing that the centroidal moment of inertia about a vertical axis is 0.032 kg·m² for the tube and 0.540 kg·m² for the rod and bracket, determine (*a*) the angular velocity of the assembly after the tube has moved to end A, (*b*) the energy lost due to the plastic impact at A.

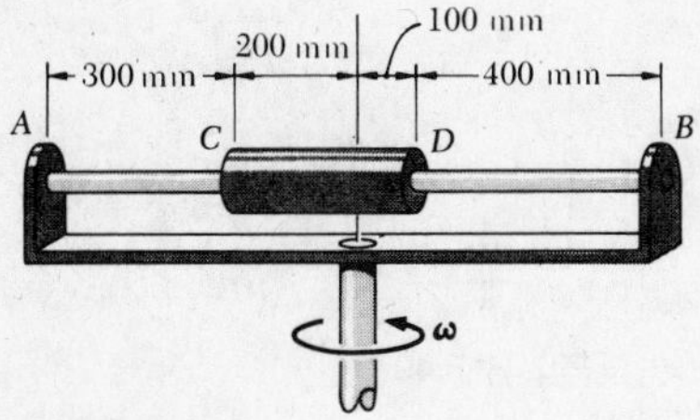

Fig. P17.42

17.43 Solve Sample Prob. 17.8, assuming that, after the cord is cut, sphere B moves to position B' but that an obstruction prevents sphere A from moving.

17.44 An 8-lb bar AB is attached by a pin at D to a 10-lb circular plate which may rotate freely about a vertical axis. Knowing that when the bar is vertical the angular velocity of the plate is 90 rpm, determine (*a*) the angular velocity of the plate after the bar has swung into a horizontal position and has come to rest against pin C, (*b*) the energy lost due to the plastic impact at C.

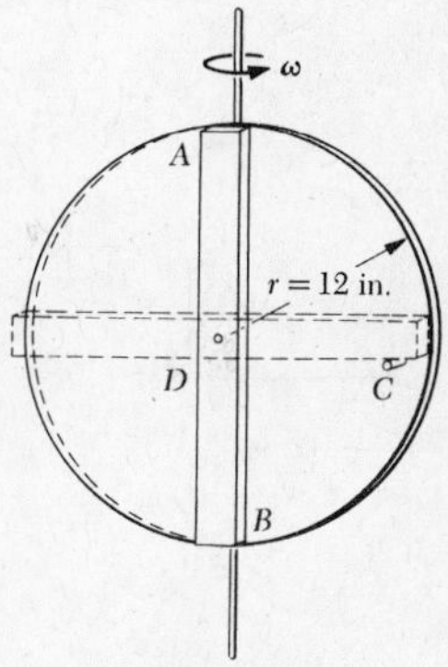

Fig. P17.44

17.45 Knowing that in Prob. 17.41 the speed of the ball is 1.2 m/s as it passes through C, determine the radial and transverse components of the velocity of the ball as it leaves the tube at B.

17.46 Collar B weighs 3 lb and may slide freely on rod OA which in turn may rotate freely in the horizontal plane. The assembly is rotating with an angular velocity $\omega = 1.5$ rad/s when a spring located between A and B is released, projecting the collar along the rod with an initial relative speed $v_r = 5$ ft/s. Knowing that the moment of inertia about O of the rod and spring is $0.15 \text{ lb} \cdot \text{ft} \cdot \text{s}^2$, determine (*a*) the minimum distance between the collar and point O in the ensuing motion, (*b*) the corresponding angular velocity of the assembly.

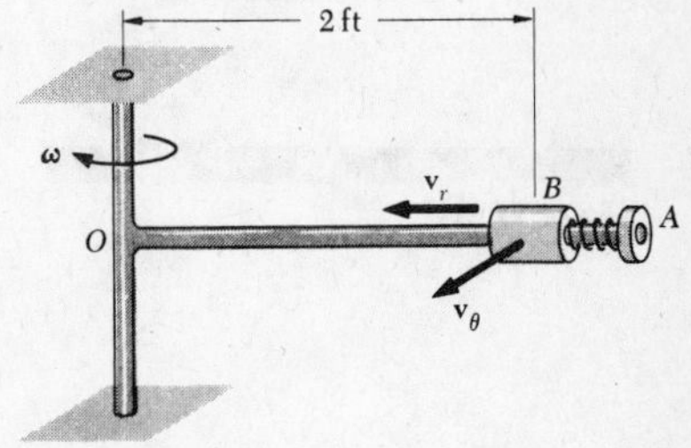

Fig. P17.46

17.47 In Prob. 17.46, determine the required magnitude of the initial relative velocity $\mathbf{v}_r$ if during the ensuing motion the minimum distance between collar B and point O is to be 1 ft.

17.48 Solve Prob. 17.46, assuming that the initial relative speed of the collar is $v_r = 10$ ft/s.

17.49 A 45-g bullet is fired with a horizontal velocity of 400 m/s into a 9-kg wooden disk of radius $R = 200$ mm. Knowing that $h = 200$ mm and that the disk is initially at rest, determine (*a*) the velocity of the center of the disk immediately after the bullet becomes embedded, (*b*) the impulsive reaction at *A*, assuming that the bullet becomes embedded in 0.001 s.

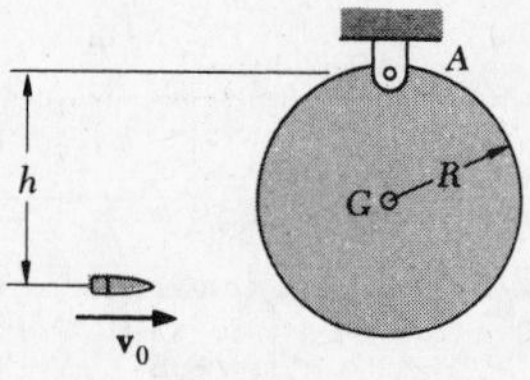

Fig. P17.49

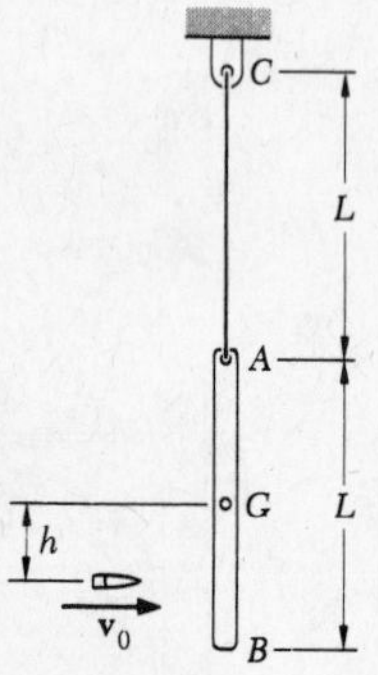

Fig. P17.51

17.50 In Prob. 17.49, determine (*a*) the required distance *h* if the impulsive reaction at *A* is to be zero, (*b*) the corresponding velocity of the center of the disk after the bullet becomes embedded.

17.51 A bullet weighing 0.08 lb is fired with a horizontal velocity of 1800 ft/s into the 15-lb wooden rod *AB* of length $L = 30$ in. The rod, which is initially at rest, is suspended by a cord of length $L = 30$ in. Knowing that $h = 6$ in., determine the velocity of each end of the rod immediately after the bullet becomes embedded.

17.52 In Prob. 17.51, determine the distance *h* for which, immediately after the bullet becomes embedded, the instantaneous center of rotation of the rod is point *C*.

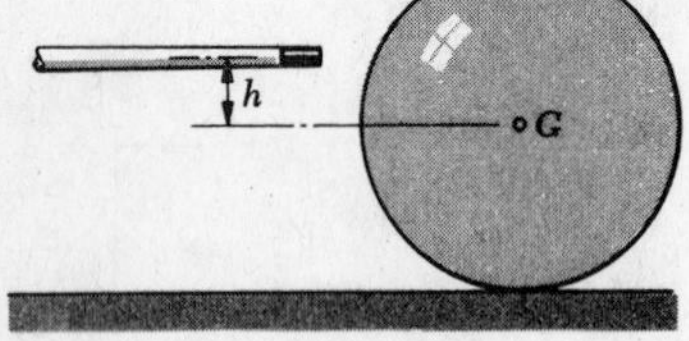

Fig. P17.53

17.53 At what height *h* above its center *G* should a billiard ball of radius *r* be struck horizontally by a cue if the ball is to start rolling without sliding?

17.54 A uniform sphere of radius *r* rolls without slipping down the incline shown. It hits the horizontal surface and, after slipping for a while, starts rolling again. Assuming that the sphere does not bounce as it hits the horizontal surface, determine its angular velocity and the velocity of its mass center after it has resumed rolling.

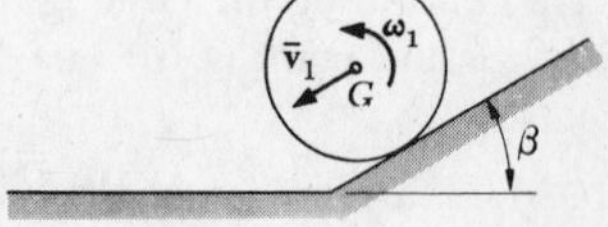

Fig. P17.54

Note to Accompany
PROBLEMS SUPPLEMENT TO VECTOR MECHANICS FOR ENGINEERS: DYNAMICS, 5TH EDITION
by Beer and Johnston

Due to a printer's error, some information in a few figures is difficult to read. While this will be corrected in future printings, the following list is supplied at this time.

Fig. P11.31 and P11.32	Collar A is on the left and collar B is on the right. Block C is suspended from the collars.
Fig. P11.82	Circle A is on the left and circle B is on the right; they both have a radius b. The distance from O to P' is r and the angle formed by OP' and OP is Ø.
Fig. P13.7	The mass of block A is 200 kg and the mass of block B is 600 kg.

17.55 A bullet of mass m is fired with a horizontal velocity v_0 and at a height $h = \frac{1}{2}R$ into a wooden disk of much larger mass M and radius R. The disk rests on a horizontal plane and the coefficient of friction between the disk and the plane is finite. (*a*) Determine the linear velocity $\bar{v}_1$ and the angular velocity ω_1 of the disk immediately after the bullet has penetrated the disk. (*b*) Describe the ensuing motion of the disk and determine its linear velocity after the motion has become uniform.

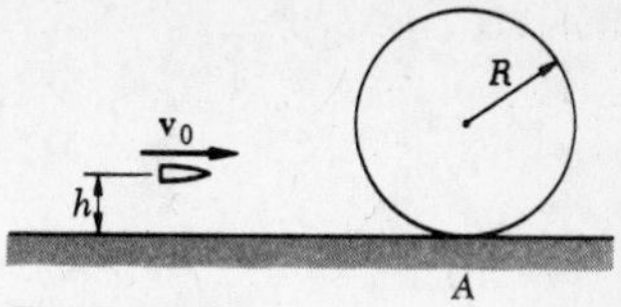

Fig. P17.55

17.56 Determine the height h at which the bullet of Prob. 17.55 should be fired (*a*) if the disk is to roll without sliding immediately after impact, (*b*) if the disk is to slide without rolling immediately after impact.

17.57 The uniform plate $ABCD$ is falling with a velocity v_1 when wire BE becomes taut. Assuming that the impact is perfectly plastic, determine the angular velocity of the plate and the velocity of its mass center immediately after the impact.

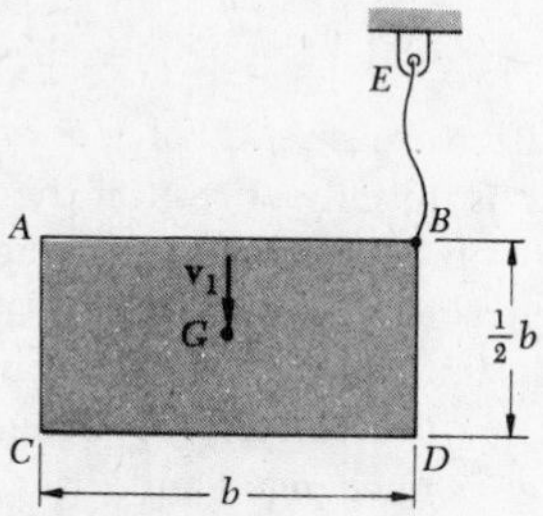

Fig. P17.57

17.58 A uniformly loaded rectangular crate is released from rest in the position shown. Assuming that the floor is sufficiently rough to prevent slipping and that the impact at B is perfectly plastic, determine the largest value of the ratio b/a for which corner A will remain in contact with the floor.

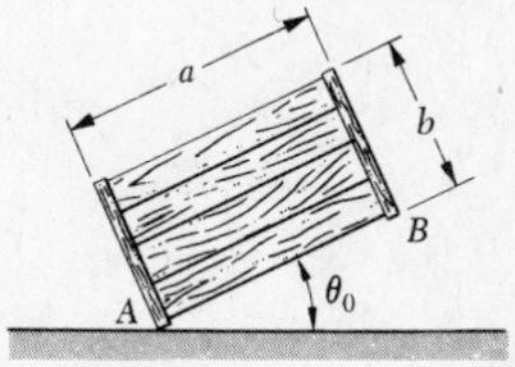

Fig. P17.58

17.59 A sphere of weight W_S is dropped from a height h and strikes at A the uniform slender plank AB of weight W which is held by two inextensible cords. Knowing that the impact is perfectly plastic and that the sphere remains attached to the plank, determine the velocity of the sphere immediately after impact.

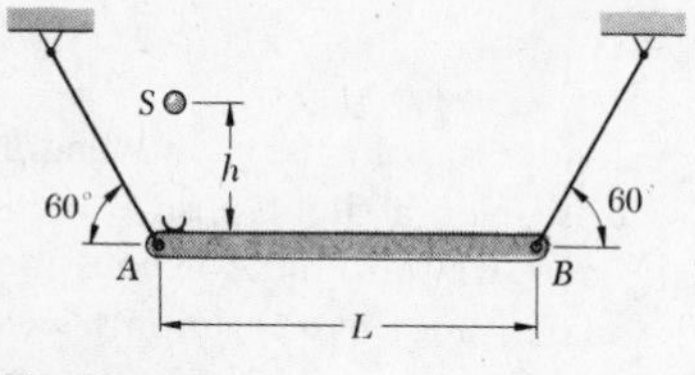

Fig. P17.59

17.60 The plank CDE of mass m_P rests on top of a small pivot at D. A gymnast A of mass m stands on the plank at end C; a second gymnast B of the same mass m jumps from a height h and strikes the plank at E. Assuming perfectly plastic impact, determine the height to which gymnast A will rise. (Assume that gymnast A stands completely rigid.)

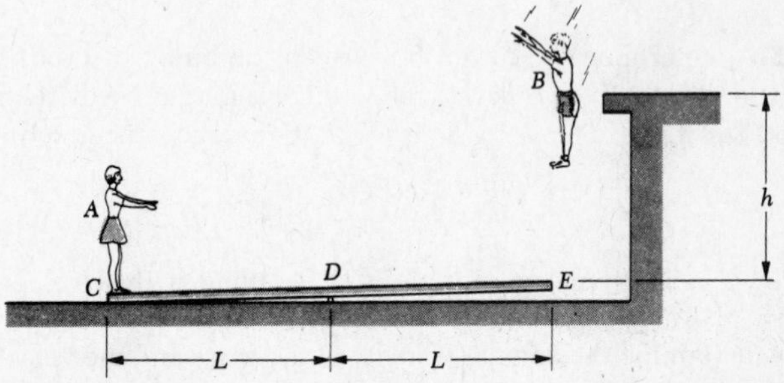

Fig. P17.60

17.61 Two uniform rods, each of mass m, form the L-shaped rigid body ABC which is initially at rest on the frictionless horizontal surface when hook D of the carriage E engages a small pin at C. Knowing that the carriage is pulled to the right with a constant velocity $\mathbf{v}_0$, determine immediately after the impact (*a*) the angular velocity of the body, (*b*) the velocity of corner B. Assume that the velocity of the carriage is unchanged and that the impact is perfectly plastic.

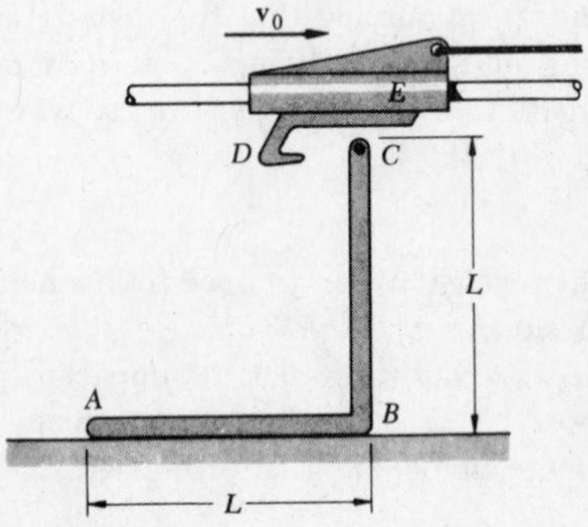

Fig. P17.61

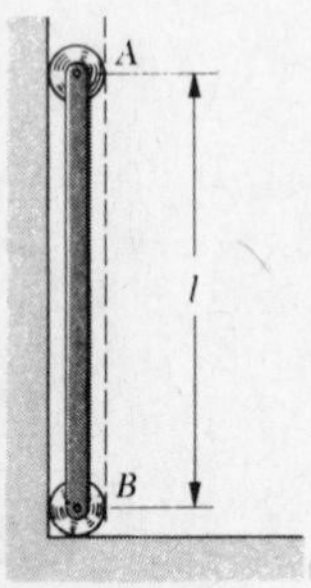

Fig. P17.62

17.62 A slender rod of mass m and length l is held in the position shown. Roller B is given a slight push to the right and moves along the horizontal plane, while roller A is constrained to move vertically. Determine the magnitudes of the impulses exerted on the rollers A and B as roller A strikes the ground. Assume perfectly plastic impact.

17.63 Two identical slender rods may swing freely from the pivots shown. Rod A is released from rest in a horizontal position and swings to a vertical position, at which time the small knob K strikes rod B which was at rest. If $h = \frac{1}{2}l$ and $e = \frac{1}{2}$, determine (a) the angle through which rod B will swing, (b) the angle through which rod A will rebound.

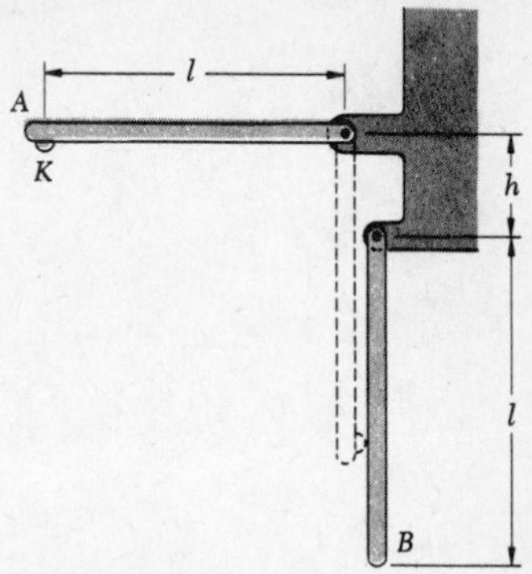

Fig. P17.63

17.64 Solve Prob. 17.63, assuming $e = 1$.

17.65 A slender rigid rod of mass m and with a centroidal radius of gyration $\bar{k}$ strikes a rigid knob at A with a vertical velocity of magnitude $\bar{v}_1$ and no angular velocity. Assuming that the impact is perfectly elastic, show that the magnitude of the velocity of G after impact is

$$\bar{v}_2 = \bar{v}_1 \frac{r^2 - \bar{k}^2}{r^2 + \bar{k}^2}$$

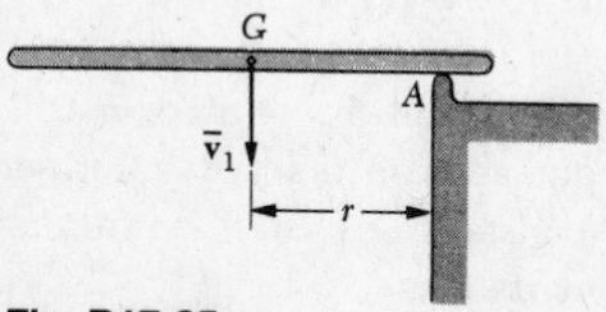

Fig. P17.65

17.66 A square block of mass m is falling with a velocity $\bar{v}_1$ when it strikes a small obstruction at B. Assuming that the impact between corner A and the obstruction B is perfectly elastic, determine the angular velocity of the block and the velocity of its mass center G immediately after impact.

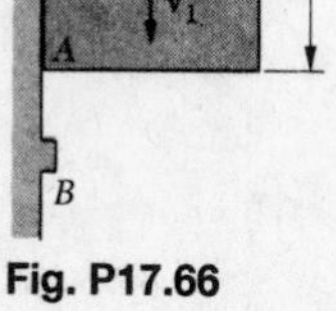

Fig. P17.66

17.67 Solve Prob. 17.60, assuming that the impact between gymnast B and the plank is perfectly elastic.

CHAPTER 18
KINETICS OF RIGID BODIES IN THREE DIMENSIONS

SECTIONS 18.1 to 18.4

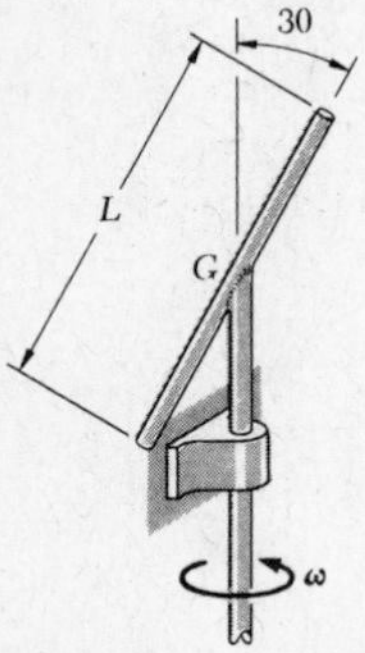

Fig. P18.1

18.1 A thin homogeneous rod of mass m and length L rotates with a constant angular velocity ω about a vertical axis through its mass center G. Determine the magnitude and direction of the angular momentum $\mathbf{H}_G$ of the rod about its mass center.

18.2 A thin homogeneous disk of mass m and radius r spins at the constant rate ω_2 about an axle held by a fork-ended horizontal rod which rotates at the constant rate ω_1. Determine the angular momentum of the disk about its mass center.

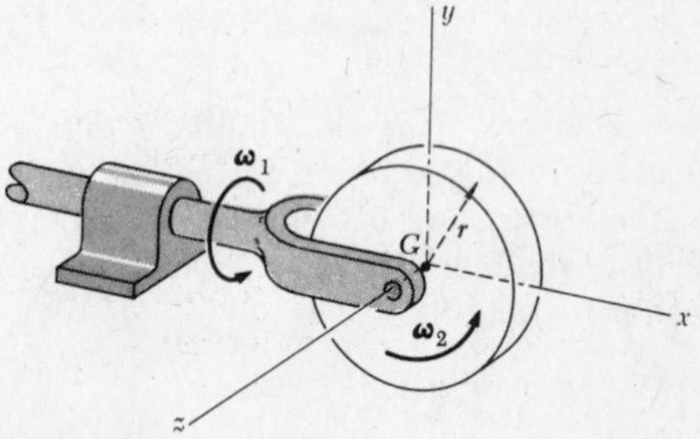

Fig. P18.2

18.3 A thin rectangular plate of weight 15 lb is attached to a shaft as shown. If the angular velocity ω of the plate is 4 radians/sec at the instant shown, determine its angular momentum about its mass center G.

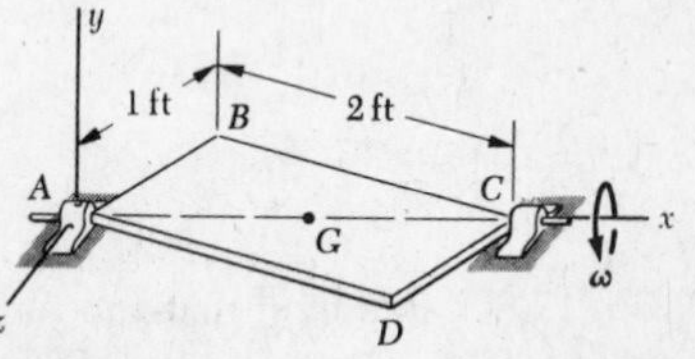

Fig. P18.3

18.4 A thin homogeneous disk of mass 800 g and radius 100 mm rotates at a constant rate $\omega_2 = 20$ rad/s with respect to the arm ABC, which itself rotates at a constant rate $\omega_1 = 10$ rad/s about the x axis. Determine the angular momentum of the disk about point D.

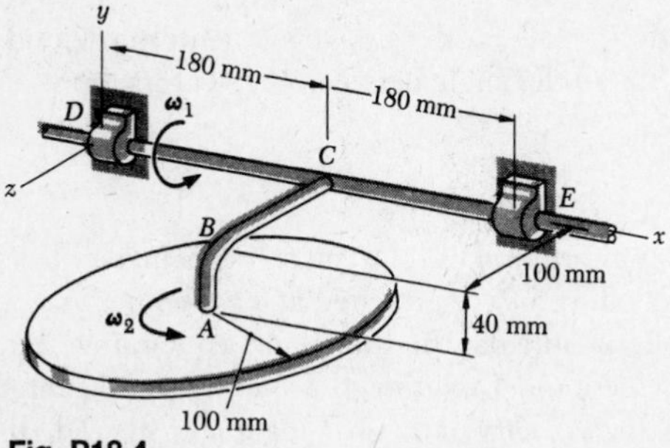

Fig. P18.4

18.5 A thin homogeneous triangular plate of mass 5 kg is welded to a light axle which can rotate freely in bearings at A and B. Knowing that the plate rotates at a constant rate $\omega = 5$ rad/s, determine its angular momentum about A.

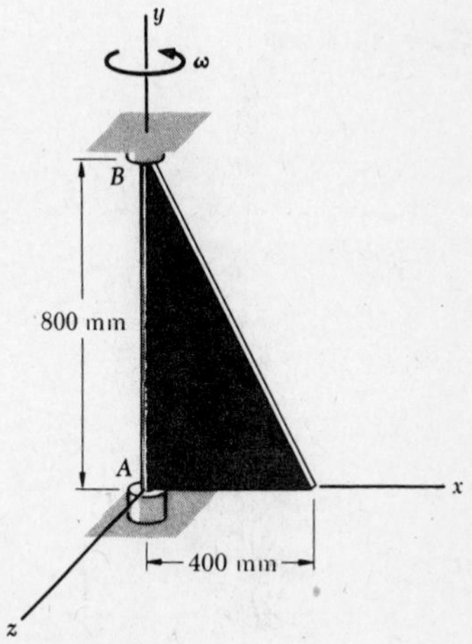

Fig. P18.5

18.6 Determine the angular momentum of the plate of Prob. 18.5 about its mass center.

18.7 The solid rectangular parallelepiped shown weighs 30 lb. Determine its angular momentum about O when it rotates about the diagonal OB at the rate $\omega = 20$ radians/sec.

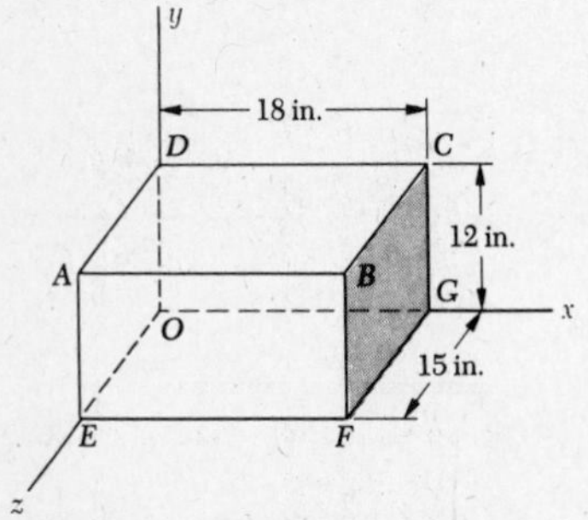

Fig. P18.7 and P18.8

18.8 The solid rectangular parallelepiped shown weighs 30 lb. Determine its angular momentum about O when it rotates about the diagonal OC of one of its faces at the rate $\omega = 20$ radians/sec.

18.9 Two L-shaped arms, each of mass 3 kg, are welded at the third points of the 900-mm shaft AB. Knowing that shaft AB rotates at a constant rate $\omega = 300$ rpm, determine (a) the angular momentum of the body about A, (b) the angle formed by the angular momentum and the shaft AB.

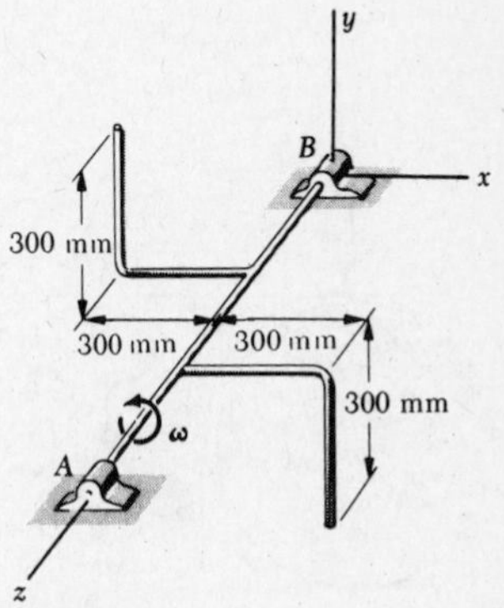

Fig. P18.9

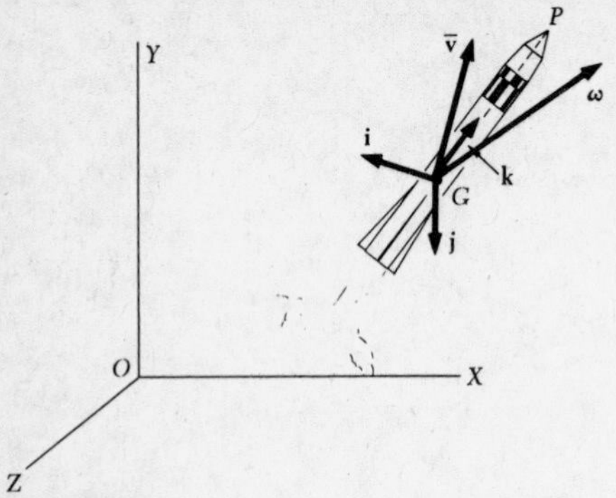

Fig. P18.10

18.10 At a given instant during its flight, a launch vehicle has an angular velocity $\boldsymbol{\omega} = (0.3 \text{ rad/s})\mathbf{j} + (2 \text{ rad/s})\mathbf{k}$ and its mass center G has a velocity $\mathbf{v} = (6 \text{ m/s})\mathbf{i} + (9 \text{ m/s})\mathbf{j} + (1800 \text{ m/s})\mathbf{k}$, where **i**, **j**, and **k** are the unit vectors corresponding to the principal centroidal axes of inertia of the vehicle. Knowing that the vehicle has a mass of 40 Mg and that its centroidal radii of gyration are $\bar{k}_x = \bar{k}_y = 6$ m and $\bar{k}_z = 1.5$ m, determine (*a*) the linear momentum $m\bar{\mathbf{v}}$ and the angular momentum $\mathbf{H}_G$, (*b*) the angle between the vectors representing $m\bar{\mathbf{v}}$ and $\mathbf{H}_G$.

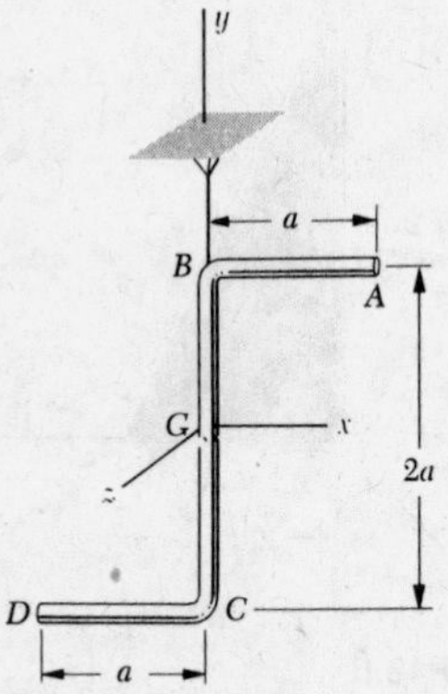

Fig. P18.11

18.11 A uniform rod of total mass m is bent into the shape shown and is suspended by a wire attached at B. The bent rod is hit at D in a direction perpendicular to the plane containing the rod (in the negative z direction). Denoting the corresponding impulse by $\mathbf{F}\Delta t$, determine (*a*) the velocity of the mass center of the rod, (*b*) the angular velocity of the rod.

18.12 Solve Prob. 18.11, assuming that the bent rod is hit at C.

18.13 A cross of total mass m, made of two rods AB and CD, each of length $2a$ and welded together at G, is suspended from a ball-and-socket joint at A. The cross is hit at C in a direction perpendicular to its plane (in the negative z direction). Denoting the corresponding impulse by $\mathbf{F}\,\Delta t$, determine immediately after impact (*a*) the angular velocity of the cross, (*b*) its instantaneous axis of rotation.

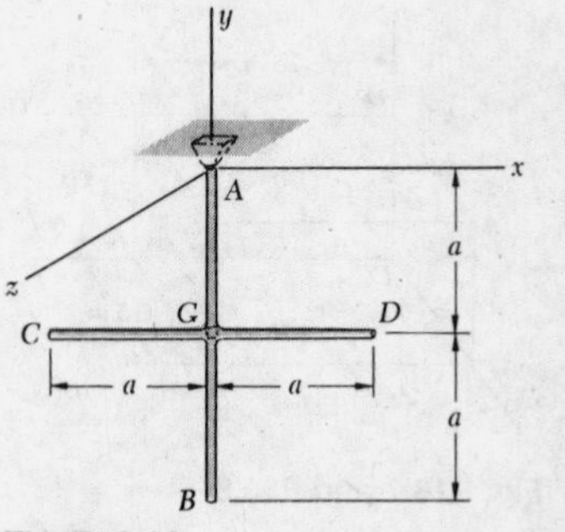

Fig. P18.13

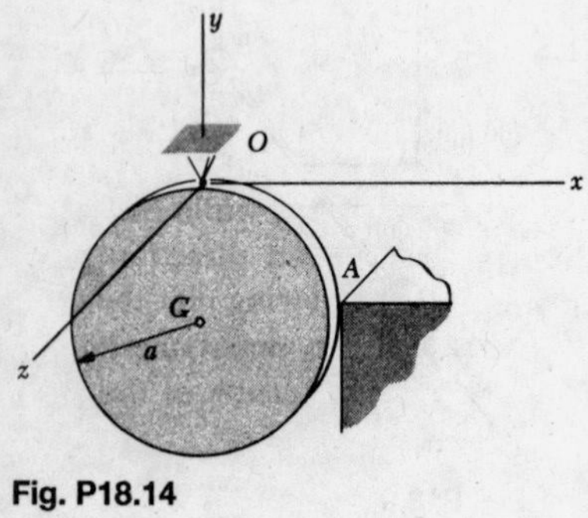

Fig. P18.14

18.14 A circular plate of radius a and mass m supported by a ball-and-socket joint at O was rotating about the x axis with a constant angular velocity $\boldsymbol{\omega} = \omega_0\mathbf{i}$ when an obstruction was suddenly introduced at A. Assuming that the impact at A is perfectly plastic, determine immediately after impact (*a*) the angular velocity of the plate, (*b*) the velocity of the mass center G.

18.15 A satellite of total weight 200 lb has no angular velocity when it is struck by a 0.02-lb meteorite traveling with a velocity $\mathbf{v}_0 = -4500\mathbf{i} + 6000\mathbf{k}$ (ft/sec) relative to the satellite. Knowing that the radii of gyration of the satellite are $\bar{k}_x = 1.50$ ft and $\bar{k}_y = \bar{k}_z = 3.00$ ft, determine the angular velocity of the satellite immediately after the meteorite has become embedded.

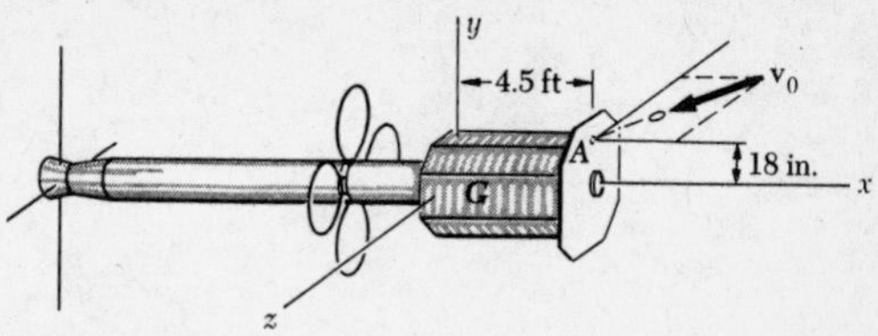

Fig. P18.15

18.16 Solve Prob. 18.15, assuming that, initially, the satellite was spinning about its axis of symmetry with an angular velocity of 10 rpm clockwise as viewed from the positive x axis.

18.17 The angular velocity of a 1000-kg space capsule is $\boldsymbol{\omega} = (0.02 \text{ rad/s})\mathbf{i} + (0.10 \text{ rad/s})\mathbf{j}$ when two small jets are activated at A and B, each in a direction parallel to the z axis. Knowing that the radii of gyration of the capsule are $\bar{k}_x = \bar{k}_z = 1.00$ m and $\bar{k}_y = 1.25$ m, and that each jet produces a thrust of 50 N, determine (*a*) the required operating time of each jet if the angular velocity of the capsule is to be reduced to zero, (*b*) the resulting change in the velocity of the mass center G.

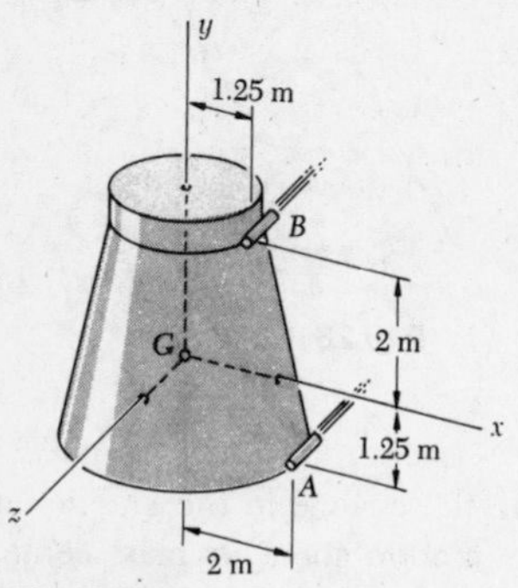

Fig. P18.17

18.18 If jet B in Prob. 18.17 is inoperative, determine (*a*) the required operating time of jet A to reduce the x component of the angular velocity $\boldsymbol{\omega}$ of the capsule to zero, (*b*) the resulting final angular velocity $\boldsymbol{\omega}$, (*c*) the resulting change in the velocity of the mass center G.

18.19 A satellite weighing 320 lb has no angular velocity when it is struck at A by a 0.04-lb meteorite traveling with a velocity $\mathbf{v}_0 = -(2400 \text{ ft/s})\mathbf{i} - (1800 \text{ ft/s})\mathbf{j} + (4000 \text{ ft/s})\mathbf{k}$ relative to the satellite. Knowing that the radii of gyration of the satellite are $\bar{k}_x = 12$ in. and $\bar{k}_y = \bar{k}_z = 16$ in., determine the angular velocity of the satellite in rpm immediately after the meteorite has become imbedded.

18.20 Solve Prob. 18.19, assuming that, initially, the satellite was spinning about its axis of symmetry with an angular velocity of 12 rpm clockwise as viewed from the positive x axis.

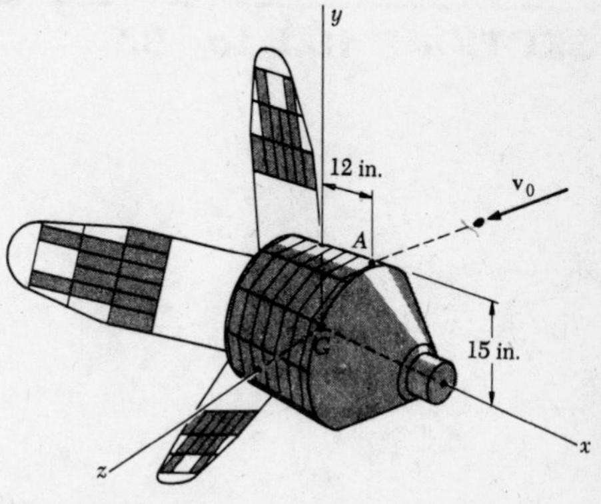

Fig. P18.19

18.21 Determine the kinetic energy of the plate of Prob. 18.3.

18.22 Determine the change in kinetic energy of the plate of Prob. 18.14 due to its impact with the obstruction.

18.23 The 400-kg space capsule is spinning with an angular velocity $\omega_0 = (100 \text{ rpm})\mathbf{k}$ when a 500-g projectile is fired from A in a direction parallel to the x axis and with a velocity $\mathbf{v}_0$ of 1200 m/s. Knowing that the radii of gyration of the capsule are $\bar{k}_x = \bar{k}_y = 450$ mm and $\bar{k}_z = 600$ mm, determine immediately after the projectile has been fired (*a*) the angular velocity of the capsule, (*b*) the kinetic energy of the capsule.

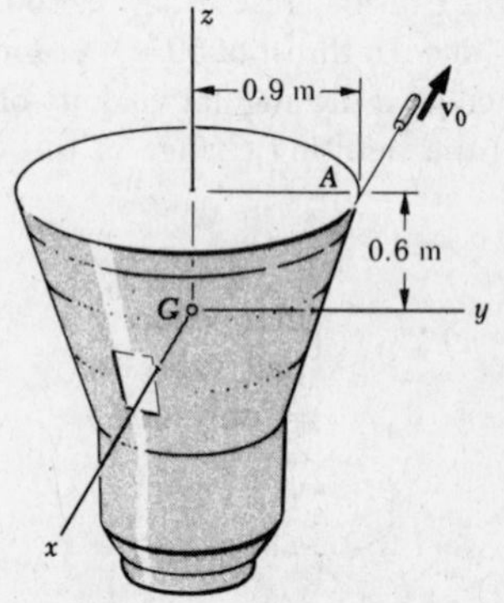

Fig. P18.23

18.24 Determine the change in the kinetic energy of the satellite of Prob. 18.15 in its motion about its mass center due to the impact of the meteorite, knowing that before the impact the satellite was spinning about its axis of symmetry with an angular velocity of 10 rpm clockwise as viewed from the positive x axis.

SECTIONS 18.5 to 18.8

18.25 Determine the rate of change $\dot{\mathbf{H}}_G$ of the angular momentum $\mathbf{H}_G$ of the disk of Prob. 18.2.

18.26 Determine the rate of change $\dot{\mathbf{H}}_A$ of the angular momentum $\mathbf{H}_A$ of the disk of Prob. 18.4.

18.27 Determine the rate of change $\dot{\mathbf{H}}_G$ of the angular momentum $\mathbf{H}_G$ of the plate of Prob. 18.3, assuming that its angular velocity remains constant.

18.28 Determine the rate of change $\dot{\mathbf{H}}_G$ of the angular momentum $\mathbf{H}_G$ of the plate of Prob. 18.3 if, at the instant considered, the angular velocity ω of the plate is 4 radians/sec and is increasing at the rate of 8 radians/sec^2.

18.29 Two 600-mm rods *BE* and *CF*, each of mass 4 kg, are attached to the shaft *AD* which rotates at a constant speed of 20 rad/s. Knowing that the two rods and the shaft lie in the same plane, determine the dynamic reactions at *A* and *D*.

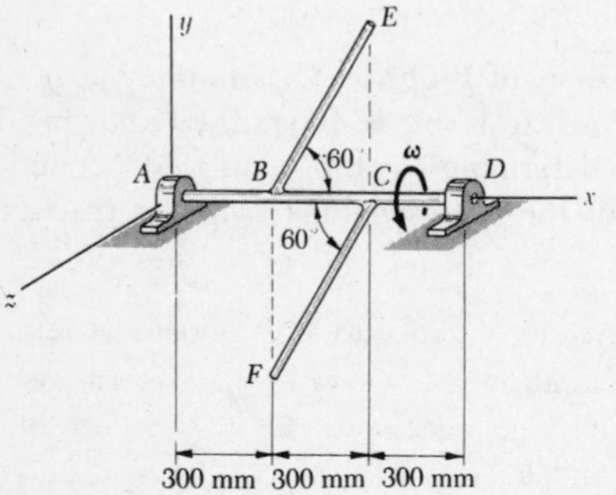

Fig. P18.29

18.30 A thin homogeneous square plate of mass m and side a is welded to a vertical shaft *AB* with which it forms an angle of 45°. Knowing that the shaft rotates with a constant angular velocity ω, determine the force-couple system representing the dynamic reaction at *A*.

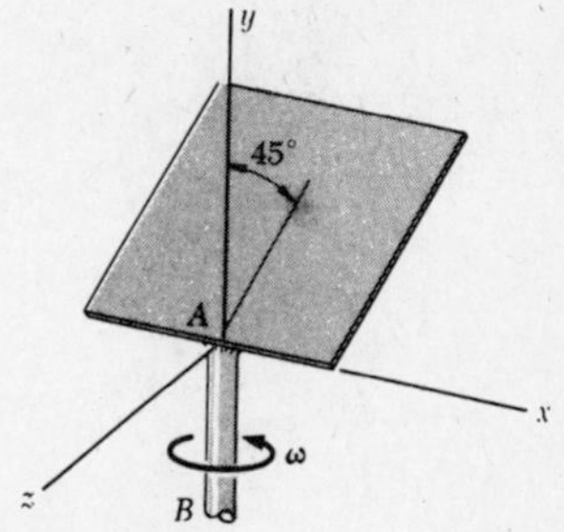

Fig. P18.30

18.31 Each element of the crankshaft shown is a homogeneous rod of weight w per unit length. Knowing that the crankshaft rotates with a constant angular velocity ω, determine the dynamic reactions at *A* and *B*.

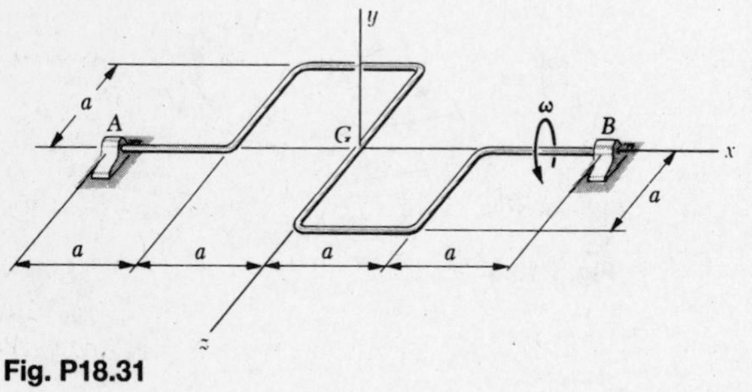

Fig. P18.31

18.32 Two triangular plates weighing 10 lb each are welded to a vertical shaft *AB*. Knowing that the system rotates at the constant rate $\omega = 6$ rad/s, determine the dynamic reactions at *A* and *B*.

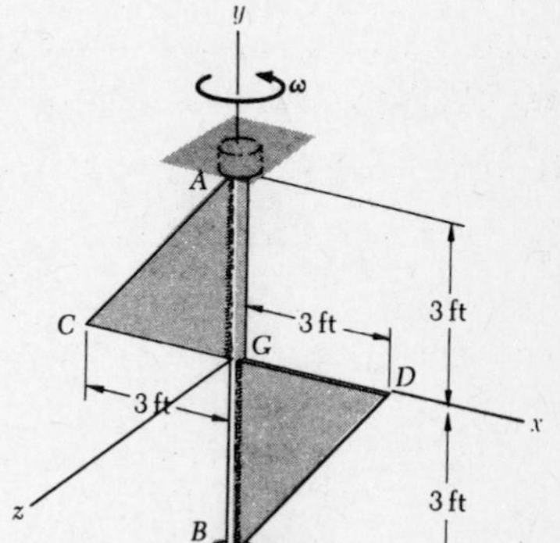

Fig. P18.32

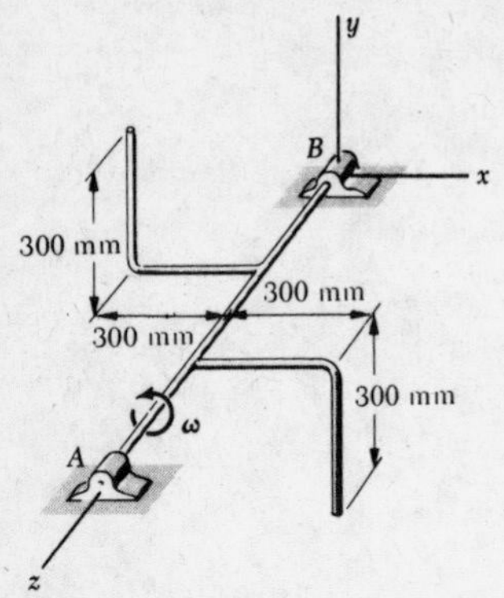

Fig. P18.33

18.33 Two L-shaped arms, each of mass 3 kg, are welded at the third points of the 900-mm shaft AB. A couple $\mathbf{M} = (20\ \text{N}\cdot\text{m})\mathbf{k}$ is applied to the shaft, which is initially at rest. Determine (*a*) the angular acceleration of the shaft, (*b*) the dynamic reactions at A and B as the shaft reaches an angular velocity of 10 rad/s.

18.34 The square plate of Prob. 18.30 is at rest ($\omega = 0$) when a couple of moment $M_0\mathbf{j}$ is applied to the shaft. Determine (*a*) the angular acceleration of the plate, (*b*) the force-couple system representing the dynamic reaction at A at that instant.

18.35 The shaft of Prob. 18.31 is initially at rest ($\omega = 0$) and is accelerated at the rate $\alpha = \dot{\omega} = 100\ \text{rad/s}^2$. Knowing that $w = 4$ lb/ft and $a = 3$ in., determine (*a*) the couple $\mathbf{M}$ required to cause the acceleration, (*b*) the corresponding dynamic reactions at A and B.

18.36 The system of Prob. 18.32 is initially at rest ($\omega = 0$) and has an angular acceleration $\boldsymbol{\alpha} = (30\ \text{rad/s}^2)\mathbf{j}$. Determine (*a*) the couple $\mathbf{M}$ required to cause the acceleration, (*b*) the corresponding dynamic reactions at A and B.

18.37 The blade of a portable saw and the rotor of its motor have a combined mass of 1.2 kg and a radius of gyration of 35 mm. Determine the couple that a man must exert on the handle to rotate the saw about the y axis with a constant angular velocity of 3 rad/s clockwise, as viewed from above, when the blade rotates at the rate $\omega = 1800$ rpm as shown.

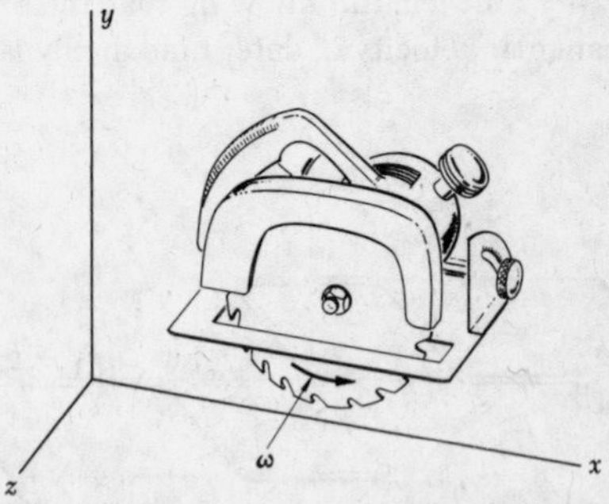

Fig. P18.37

18.38 A three-bladed airplane propeller has a mass of 120 kg and a radius of gyration of 900 mm. Knowing that the propeller rotates at 1500 rpm, determine the moment of the couple applied by the propeller to its shaft when the airplane travels in a circular path of 360-m radius at 600 km/h.

18.39 The flywheel of an automobile engine, which is mounted on the crankshaft, is equivalent to a 16-in.-diameter steel plate of $\frac{15}{16}$-in. thickness. At a time when the flywheel is rotating at 4000 rpm the automobile is traveling around a curve of 600-ft radius at a speed of 60 mi/h. Determine, at that time, the magnitude of the couple exerted by the flywheel on the horizontal crankshaft. (Specific weight of steel = 490 lb/ft^3.)

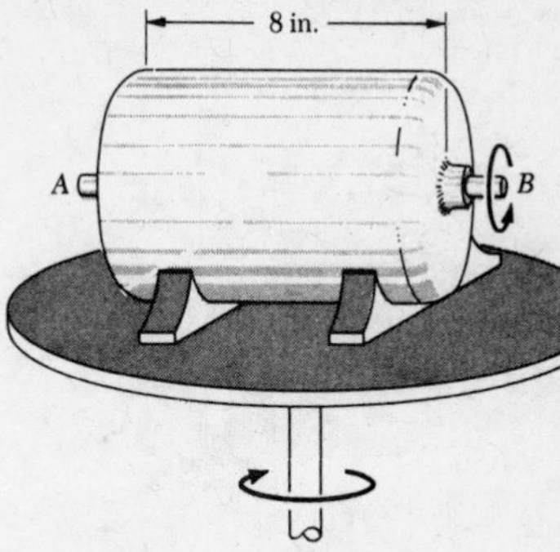

Fig. P18.40

18.40 An electric motor is supported by a rotating table. Its rotor weighs 6 lb, has a radius of gyration of 2 in., and has an angular velocity of 3600 rpm counterclockwise as viewed from the right. Determine the reactions exerted by the bearings on the axle AB when the table is rotated at the rate of 6 rpm clockwise as viewed from above.

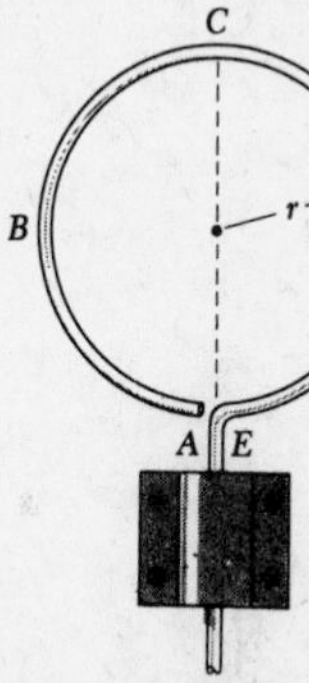

Fig. P18.41

18.41 A thin homogeneous wire, of mass m per unit length and in the shape of a circle of radius r, is made to rotate about a vertical shaft with a constant angular velocity ω. Determine the bending moment in the wire (a) at point C, (b) at point E, (c) at point B. (Neglect the effect of gravity.)

18.42 A uniform disk of radius r is welded to a rod AB of negligible weight, which is attached to the pin of a clevis which rotates with a constant angular velocity ω. Derive an expression (a) for the constant angle β that the rod forms with the vertical, (b) for the maximum value of ω for which the rod will remain vertical ($\beta = 0$).

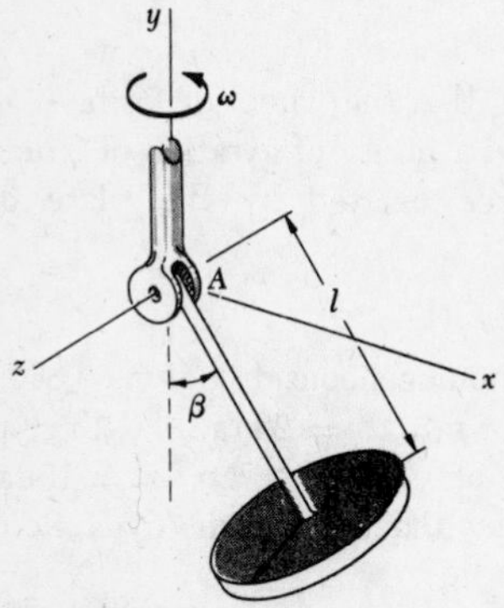

Fig. P18.42

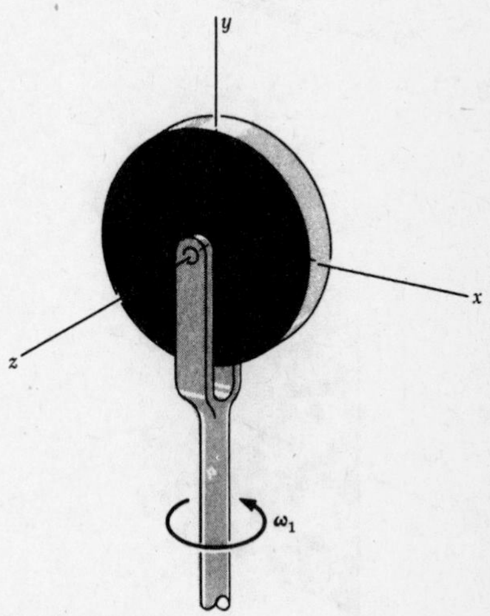

Fig. P18.43

18.43 A thin homogeneous disk of mass m and radius r spins at the constant rate ω_2 about a horizontal axle held by a fork-ended vertical rod which rotates at the constant rate ω_1. Determine the couple $\mathbf{M}$ exerted by the rod on the disk.

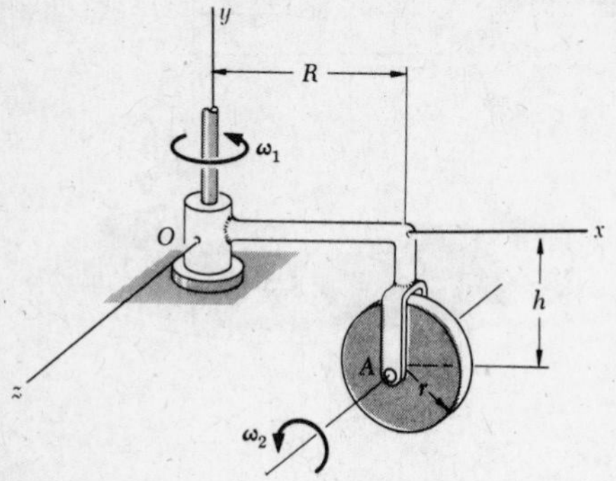

Fig. P18.44

18.44 A disk of mass m and radius r rotates at a constant rate ω_2 with respect to the arm OA, which itself rotates at a constant rate ω_1 about the y axis. Determine the force-couple system representing the dynamic reaction at O.

18.45 A stationary horizontal plate is attached to the ceiling by means of a fixed vertical tube. A wheel of radius a and mass m is mounted on a light axle AC which is attached by means of a clevis at A to a rod AB fitted inside the vertical tube. The rod AB is made to rotate with a constant angular velocity Ω causing the wheel to roll on the lower face of the stationary plate. Determine the minimum angular velocity Ω for which contact is maintained between the wheel and the plate. Consider the particular cases (*a*) when the mass of the wheel is concentrated in the rim, (*b*) when the wheel is equivalent to a thin disk of radius a.

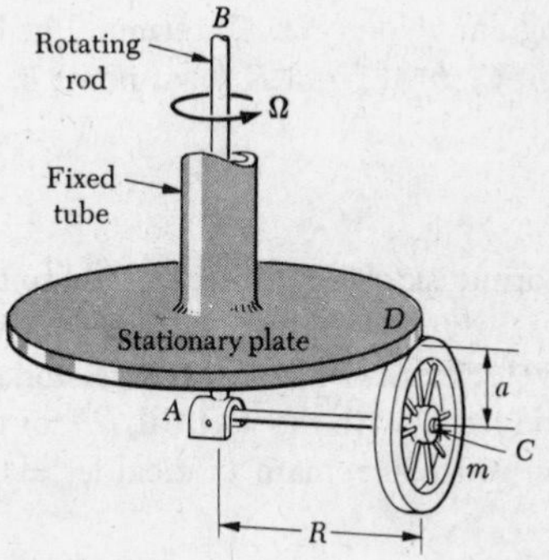

Fig. P18.45

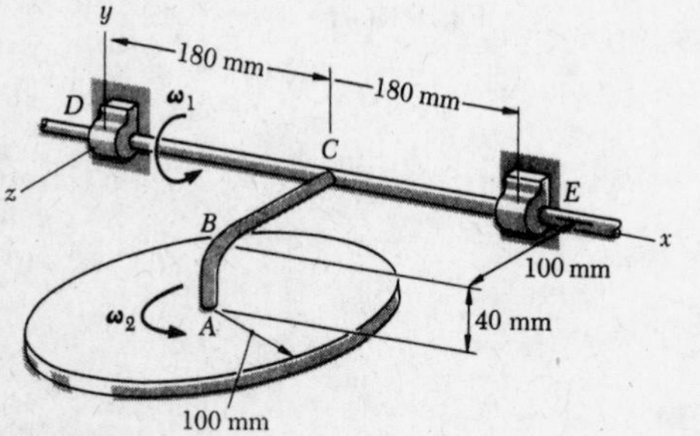

Fig. P18.47

18.46 Assuming that the wheel of Prob. 18.45 weighs 8 lb, has a radius $a = 4$ in. and a radius of gyration of 3 in., and that $R = 20$ in., determine the force exerted by the plate on the wheel when $\Omega = 25$ rad/s.

18.47 A thin homogeneous disk of mass 800 g and radius 100 mm rotates at a constant rate $\omega_2 = 20$ rad/s with respect to the arm ABC, which itself rotates at a constant rate $\omega_1 = 10$ rad/s about the x axis. For the position shown, determine the dynamic reactions at the bearings D and E.

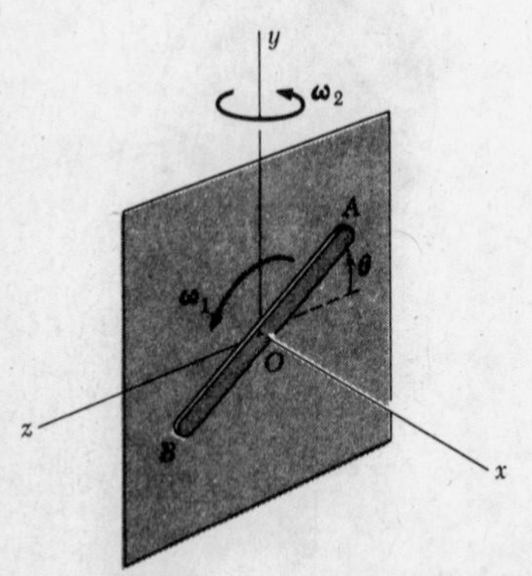

Fig. P18.48

18.48 A slender homogeneous rod AB of mass m and length L is made to rotate at the constant rate ω_1 about the horizontal x axis, while the vertical plane in which it rotates is made to rotate at the constant rate ω_2 about the vertical y axis. Express as a function of the angle θ (*a*) the couple $M_1\mathbf{i}$ required to maintain the rotation of the rod in the vertical plane, (*b*) the couple $M_2\,\mathbf{j}$ required to maintain the rotation of that plane.

18.49 The rate of steady precession $\dot{\phi}$ of the cone shown about the vertical is observed to be 30 rpm. Knowing that $r = 75$ mm and $h = 300$ mm, determine the rate of spin $\dot{\psi}$ of the cone about its axis of symmetry if $\beta = 120°$.

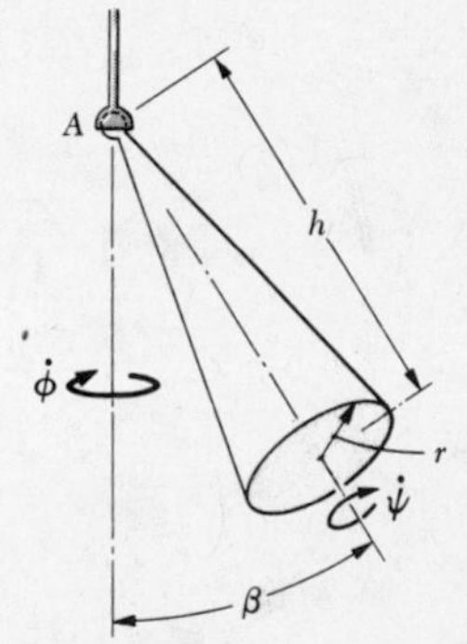

Fig. P18.49

18.50 Solve Prob. 18.49, assuming the same rate of steady precession and $\beta = 60°$.

18.51 A high-speed photographic record shows that a certain projectile was fired with a horizontal velocity $\bar{\mathbf{v}}$ of 2000 ft/s and with its axis of symmetry forming an angle $\beta = 3°$ with the horizontal. The rate of spin $\dot{\psi}$ of the projectile was 6000 rpm, and the atmospheric drag was equivalent to a force **D** of 25 lb acting at the center of pressure C_P located at a distance $c = 3$ in. from G. (*a*) Knowing that the projectile weighs 40 lb and has a radius of gyration of 2 in. with respect to its axis of symmetry, determine its approximate rate of steady precession. (*b*) If it is further known that the radius of gyration of the projectile with respect to a transverse axis through G is 8 in., determine the exact values of the two possible rates of precession.

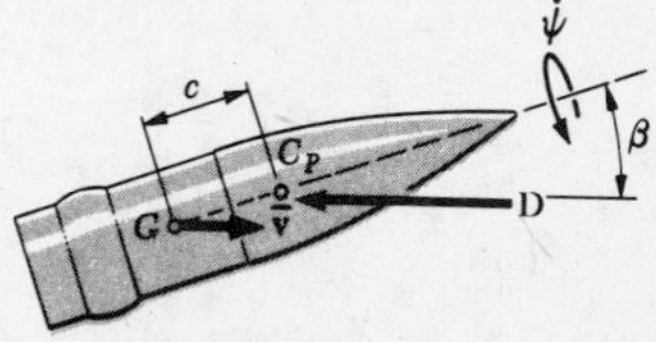

Fig. P18.51

18.52 Determine the precession axis and the rates of precession and spin of a rod which is given an initial angular velocity ω of 12 rad/s in the direction shown.

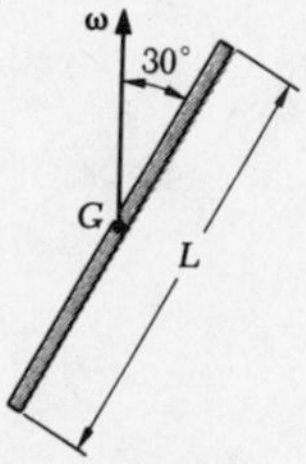

Fig. P18.52

18.53 Determine the precession axis and the rates of precession and spin of the satellite of Prob. 18.19 after the impact.

18.54 Determine the precession axis and the rates of precession and spin of the satellite of Prob. 18.19 knowing that, before impact, the angular velocity of the satellite was $\omega_0 = -(12 \text{ rpm})\mathbf{i}$.

18.55 The space capsule has no angular velocity when the jet at A is activated for 1 s in a direction parallel to the x axis. Knowing that the capsule has a mass of 1000 kg, that its radii of gyration are $\bar{k}_x = \bar{k}_y = 1.00$ m and $\bar{k}_z = 1.25$ m, and that the jet at A produces a thrust of 50 N, determine the axis of precession and the rates of precession and spin after the jet has stopped.

18.56 The space capsule has an angular velocity $\omega = (0.02 \text{ rad/s})\mathbf{j} + (0.10 \text{ rad/s})\mathbf{k}$ when the jet at B is activated for 1 s in a direction parallel to the x axis. Knowing that the capsule has a mass of 1000 kg, that its radii of gyration are $\bar{k}_x = \bar{k}_y = 1.00$ m and $\bar{k}_z = 1.25$ m, and that the jet at B produces a thrust of 50 N, determine the axis of precession and the rates of precession and spin after the jet has stopped.

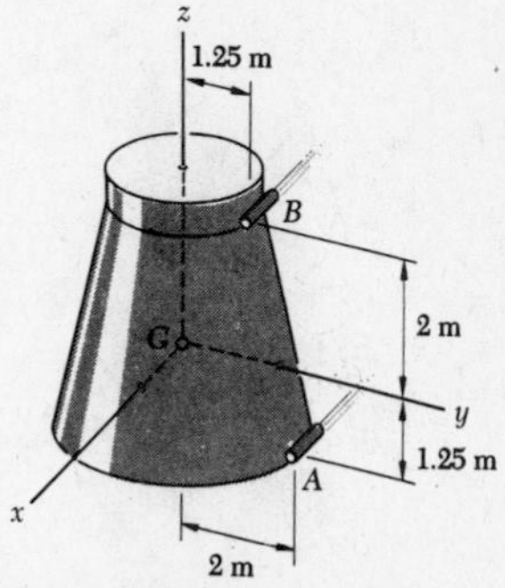

Fig. P18.55 and P18.56

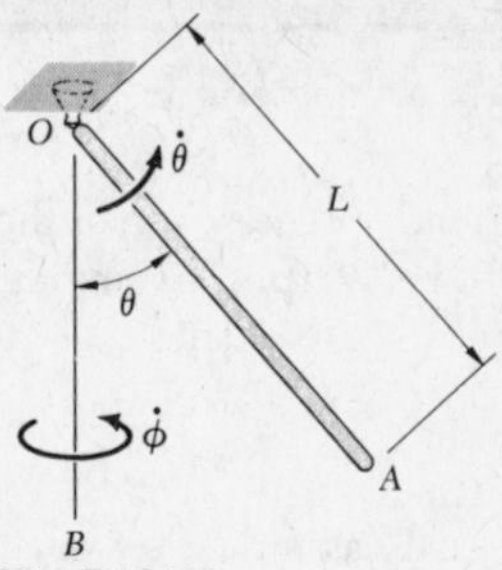

Fig. P18.57

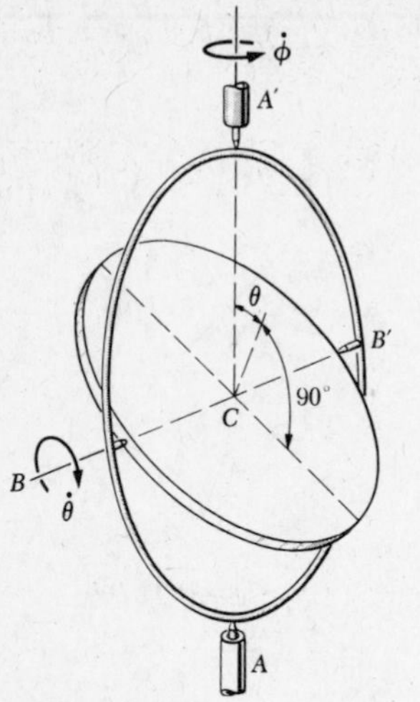

Fig. P18.58

18.57 A slender homogeneous rod OA of mass m and length L is supported by a ball-and-socket joint at O and may swing freely under its own weight. If the rod is held in a horizontal position ($\theta = 90°$), what initial angular velocity $\dot{\phi}_0$ should be given to the rod about the vertical OB if the smallest value of θ in the ensuing motion is to be 60°? (Apply the principle of conservation of energy and the principle of impulse and momentum, observing that, since $\Sigma M_{OB} = 0$, the component of $\mathbf{H}_O$ along OB must be constant.)

18.58 The gimbal $ABA'B'$ is of negligible mass and may rotate freely about the vertical AA'. The uniform disk of radius a and mass m may rotate freely about its diameter BB', which is also the horizontal diameter of the gimbal. (*a*) Applying the principle of conservation of energy, and observing that, since $\Sigma M_{AA'} = 0$, the component of the angular momentum of the disk along the fixed axis AA' must be constant, write two first-order differential equations defining the motion of the disk. (*b*) Given the initial conditions $\theta_0 \neq 0$, $\dot{\phi}_0 \neq 0$, and $\dot{\theta}_0 = 0$, express the rate of nutation $\dot{\theta}$ as a function of θ. (*c*) Show that the angle θ will never be larger than θ_0 during the ensuing motion.

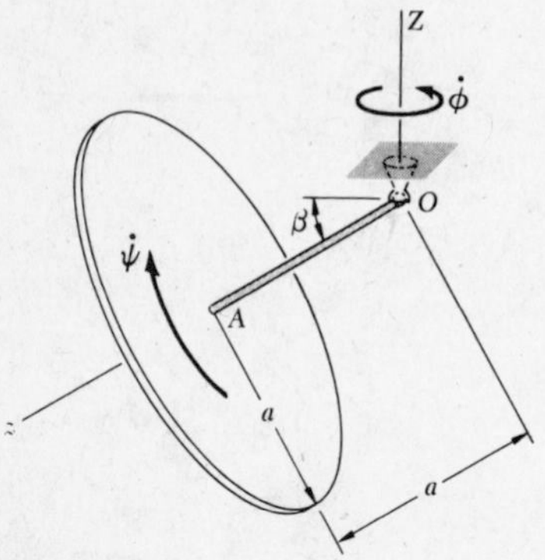

Fig. P18.59

***18.59** A thin homogeneous disk of radius a and mass m is mounted on a light axle OA of length a which is held by a ball-and-socket support at O. The disk is released in the position $\beta = 0$ with a rate of spin $\dot{\psi}_0$, clockwise as viewed from O, and with no precession or nutation. Knowing that the largest value of β in the ensuing motion is 30°, determine in terms of $\dot{\psi}_0$ the rates of precession and spin of the disk when $\beta = 30°$. (*Hint.* The angular momentum of the disk is conserved about both the Z and z axes.)

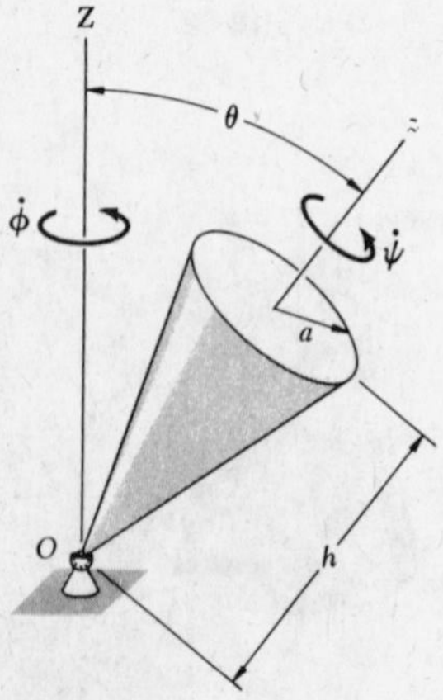

Fig. P18.60

***18.60** A solid homogeneous cone of mass m, radius a, and height $h = \frac{3}{2}a$, is held by a ball-and-socket support O. Initially the axis of symmetry of the cone is vertical ($\theta = 0$) with the cone spinning about it at the constant rate $\dot{\psi}_0$, counterclockwise as viewed from above. However, after being slightly disturbed, the cone starts falling and precessing. If the largest value of θ in the ensuing motion is 90°, determine (*a*) the rate of spin $\dot{\psi}_0$ of the cone in its initial vertical position, (*b*) the rates of precession and spin as the cone passes through its lowest position ($\theta = 90°$). (*Hint.* Use the principle of conservation of energy and the fact that the angular momentum of the cone is conserved about both the Z and z axes.)

CHAPTER 19
MECHANICAL VIBRATIONS

SECTIONS 19.1 to 19.3

19.1 The analysis of the motion of a particle shows a maximum acceleration of 30 m/s^2 and a frequency of 120 cycles per minute. Assuming that the motion is simple harmonic, determine (*a*) the amplitude, (*b*) the maximum velocity.

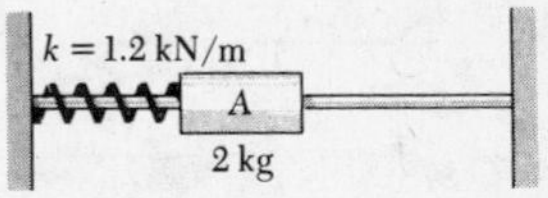

Fig. P19.2

19.2 Collar *A* is attached to the spring shown and may slide without friction on the horizontal rod. If the collar is moved 75 mm from its equilibrium position and released, determine the period, the maximum velocity, and the maximum acceleration of the resulting motion.

19.3 A particle moves in simple harmonic motion with an amplitude of 4 in. and a period of 0.60 s. Find the maximum velocity and the maximum acceleration.

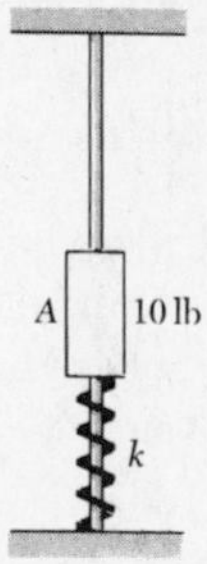

Fig. P19.4

19.4 The 10-lb collar is attached to a spring of constant $k = 4$ lb/in. as shown. If the collar is given a displacement of 2 in. from its equilibrium position and released, determine for the ensuing motion (*a*) the period, (*b*) the maximum velocity of the collar, (*c*) the maximum acceleration of the collar.

19.5 In Prob. 19.4, determine the position, velocity, and acceleration of the collar 0.20 sec after it has been released.

19.6 In Prob. 19.4, determine (*a*) the time required for the collar to move 3 in. upward after it has been released, (*b*) the corresponding velocity and acceleration of the collar.

19.7 and 19.8 A 35-kg block is supported by the spring arrangement shown. If the block is moved vertically downward from its equilibrium position and released, determine (*a*) the period and frequency of the resulting motion, (*b*) the maximum velocity and acceleration of the block if the amplitude of the motion is 20 mm.

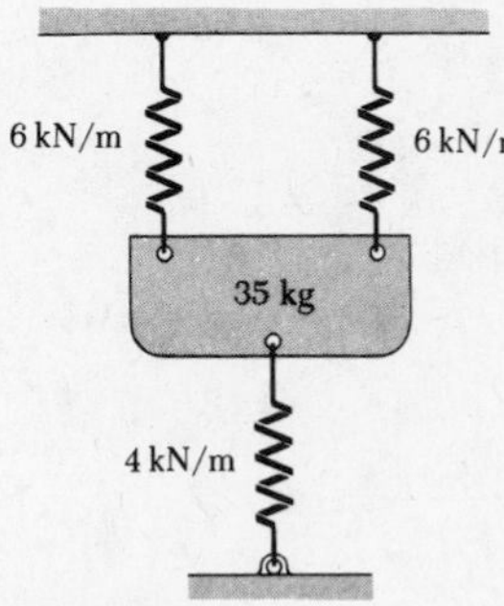

Fig. P19.7

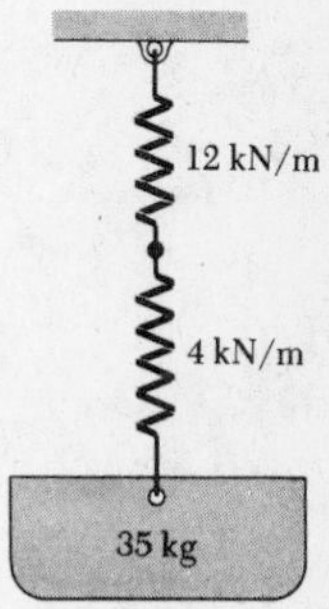

Fig. P19.8

19.9 A simple pendulum of length l is suspended in an elevator. A mass m is attached to a spring of constant k and is carried in the same elevator. Determine the period of vibration of both the pendulum and the mass if the elevator has an upward acceleration **a**.

19.10 Determine (*a*) the required length l of a simple pendulum if the period of small oscillations is to be 2 s, (*b*) the required amplitude of this pendulum if the maximum velocity of the bob is to be 200 mm/s.

Fig. P19.11

19.11 The frequency of vibration of the system shown is observed to be 2.5 cycles per second. After cylinder B is removed, the frequency is observed to be 3 cycles per second. Knowing that cylinder B weighs 2 lb, determine the weight of cylinder A.

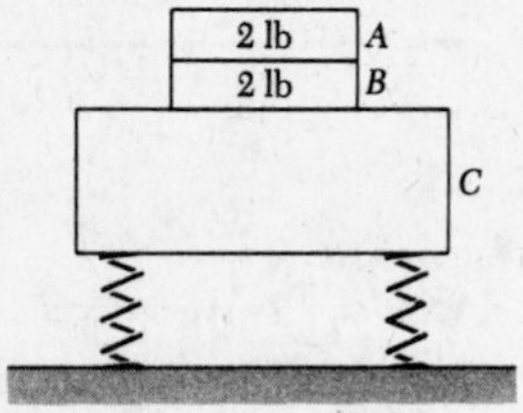

Fig. P19.12

19.12 The period of vibration of the system shown is observed to be 0.8 s. If block A is removed, the period is observed to be 0.7 s. Determine (*a*) the weight of block C, (*b*) the period of vibration when both blocks A and B have been removed.

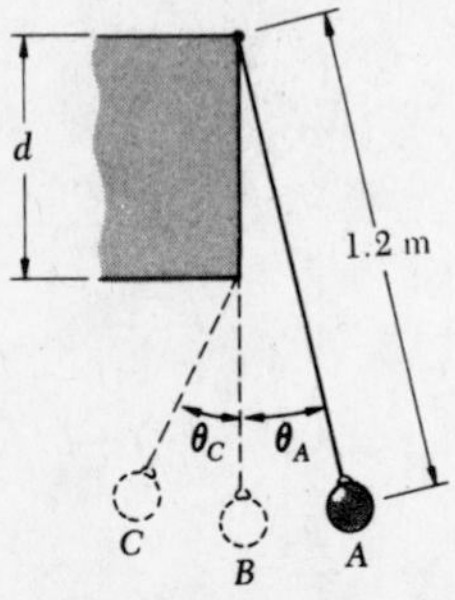

Fig. P19.13 and P19.14

19.13 A small bob is attached to a cord of length 1.2 m and is released from rest when $\theta_A = 5°$. Knowing that $d = 0.6$ m, determine (*a*) the time required for the bob to return to point A, (*b*) the amplitude θ_C.

19.14 A small bob is attached to a cord of length 1.2 m and released from rest at A when $\theta_A = 4°$. Determine the distance d for which the bob will return to point A in 2 s.

19.15 and 19.16 An 80-lb block is attached to the rod AB and to the two springs shown. Knowing that $h = 30$ in., determine the period of small oscillations of the block. Neglect the weight of the rod and assume that each spring can act in either tension or compression.

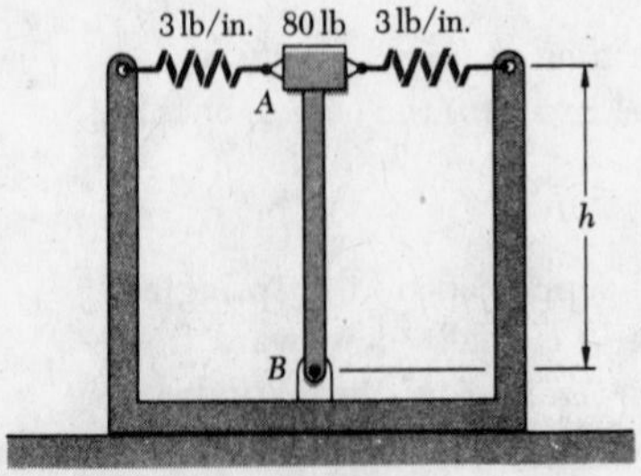

Fig. P19.15

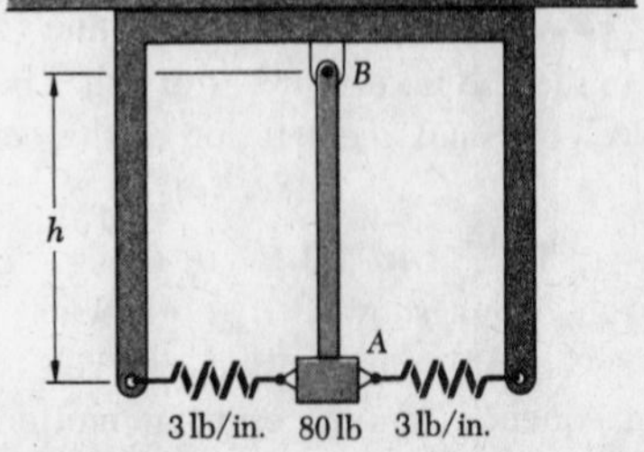

Fig. P19.16

SECTION 19.5

19.17 and 19.18 The 4-kg uniform rod shown is attached to a spring of constant $k = 450$ N/m. If end A of the rod is depressed 50 mm and released, determine (*a*) the period of vibration, (*b*) the maximum velocity of end A.

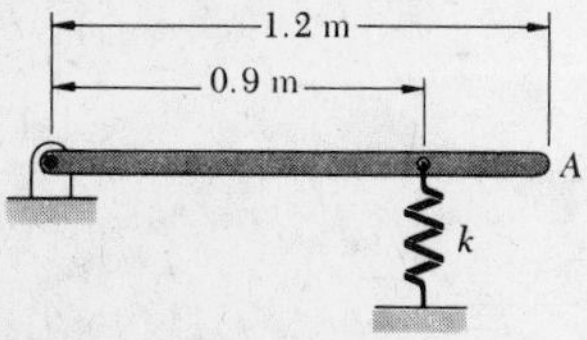

Fig. P19.17

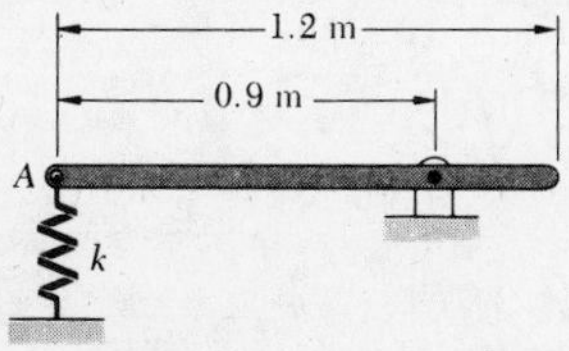

Fig. P19.18

19.19 A belt is placed over the rim of a 30-lb disk as shown and then attached to a 10-lb weight and to a spring of constant $k = 3$ lb/in. If the weight is moved 2 in. down from its equilibrium position and released, determine (*a*) the period of vibration, (*b*) the maximum velocity of the weight. Assume friction is sufficient to prevent the belt from slipping on the rim.

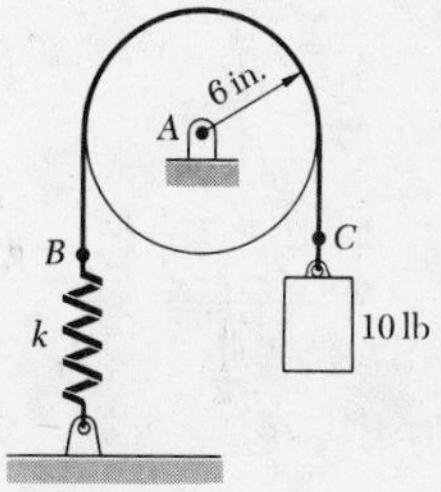

Fig. P19.19

19.20 In Prob. 19.19, determine (*a*) the frequency of vibration, (*b*) the maximum tension which occurs in the belt at B and C.

19.21 A uniform square plate of mass m is supported in a horizontal plane by a vertical pin at B and is attached at A to a spring of constant k. If corner A is given a small displacement and released, determine the period of the resulting motion.

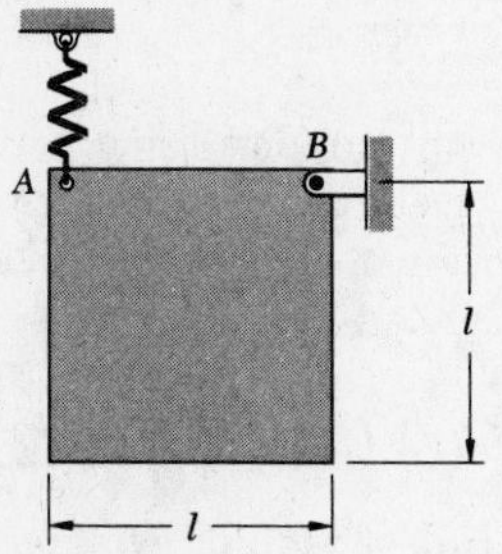

Fig. P19.21

19.22 A semicircular rod of mass m is supported in a horizontal plane by a vertical pin at A and is attached at B to a spring of constant k. If end B is given a small displacement and released, determine the frequency of the resulting motion.

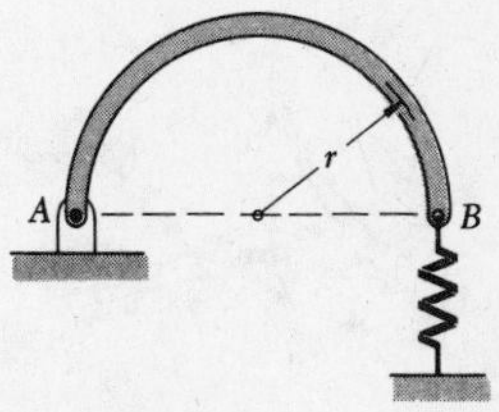

Fig. P19.22

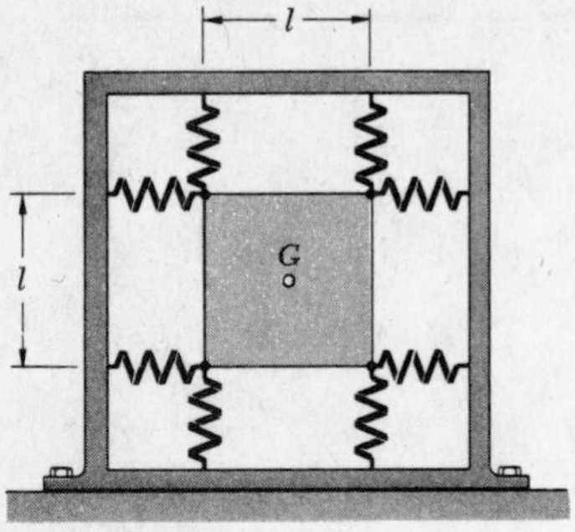

Fig. P19.23

19.23 A square plate of mass m is held by eight springs, each of constant k. Knowing that each spring can act in either tension or compression, determine the frequency of the resulting vibration (*a*) if the plate is given a small vertical displacement and released, (*b*) if the plate is rotated through a small angle about G and released.

19.24 A uniform rod of mass m is supported by a pin at its midpoint C and is attached to a spring of constant k. If end A is given a small displacement and released, determine the frequency of the resulting motion.

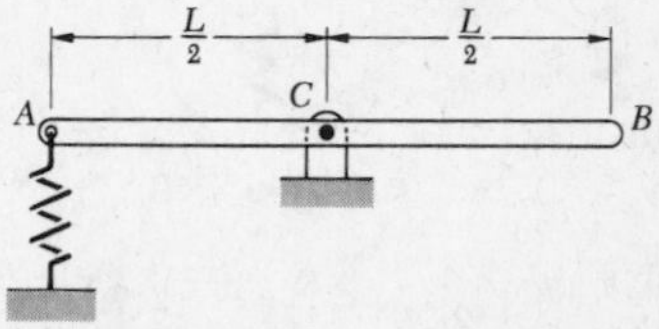

Fig. P19.24

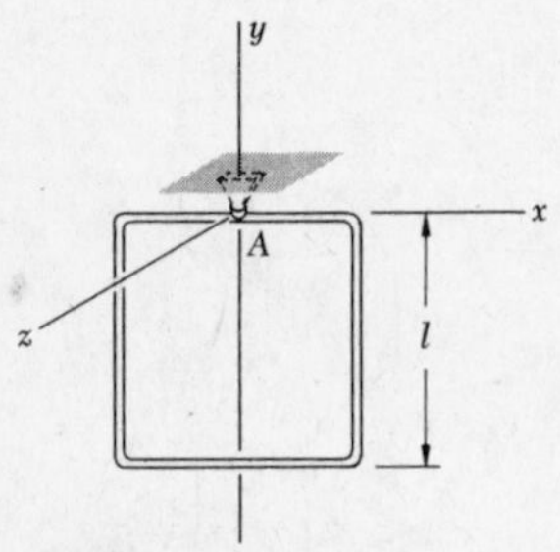

Fig. P19.25

19.25 A homogeneous wire is bent to form a square of side l which is supported by a ball-and-socket joint at A. Determine the period of small oscillations of the square (*a*) in the plane of the square, (*b*) in a direction perpendicular to the square.

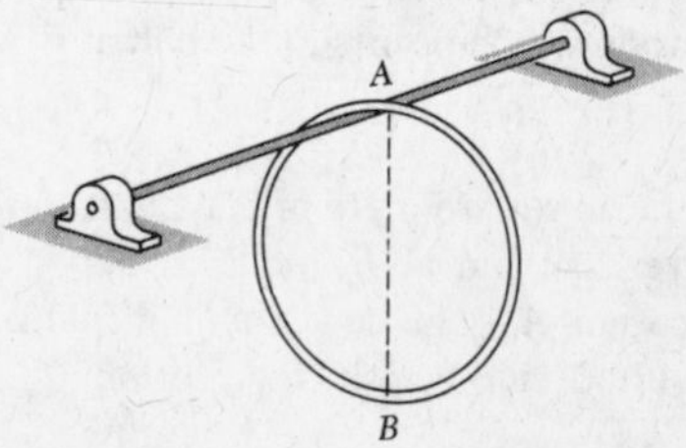

Fig. P19.26

19.26 A thin hoop of radius r and mass m is suspended from a rough rod as shown. Determine the frequency of small oscillations of the hoop (*a*) in the plane of the hoop, (*b*) in a direction perpendicular to the plane of the hoop. Assume that μ is sufficiently large to prevent slipping at A.

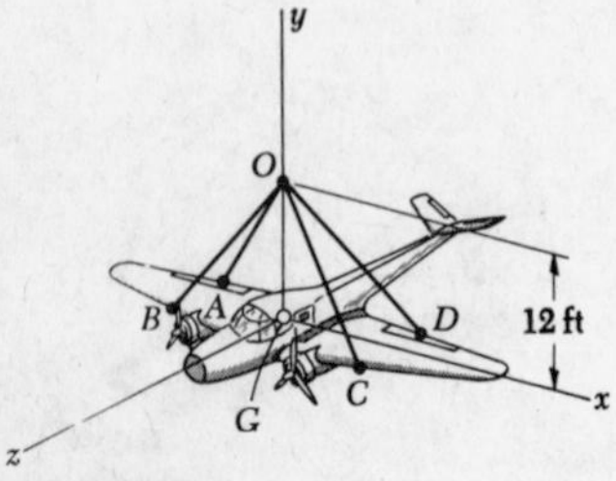

Fig. P19.27

19.27 The centroidal radii of gyration $\bar{k}_x$ and $\bar{k}_z$ of an airplane are determined by letting the airplane oscillate as a compound pendulum. In the arrangement shown, the distance from the mass center G to the point of suspension O is known to be 12 ft. Knowing that the observed periods of oscillation about axes through O parallel to the x and z axes are, respectively, 4.3 and 4.4 s, determine the centroidal radii of gyration $\bar{k}_x$ and $\bar{k}_z$.

19.28 The centroidal radius of gyration $\bar{k}_y$ of an airplane is determined by suspending the airplane by two 12-ft-long cables as shown. The airplane is rotated through a small angle about the vertical through G and then released. Knowing that the observed period of oscillation is 3.3 s, determine the centroidal radius of gyration $\bar{k}_y$.

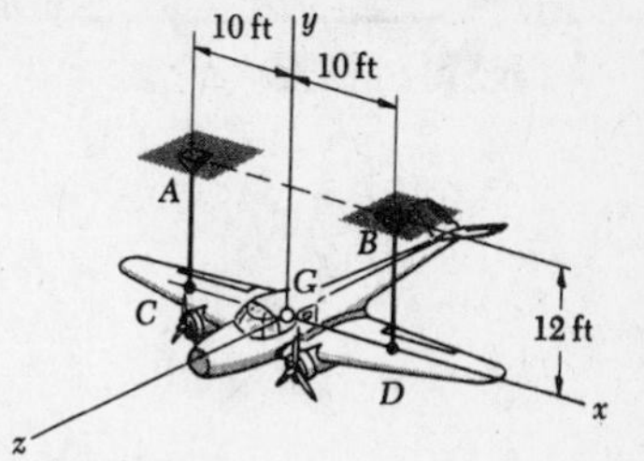

Fig. P19.28

19.29 A uniform disk having a mass of 2 kg is suspended from a steel wire which is known to have a torsional constant $K = 30$ mN·m/rad. If the disk is rotated through 360° about the vertical and then released, determine (*a*) the period of oscillation, (*b*) the maximum velocity of a point on the rim of the disk.

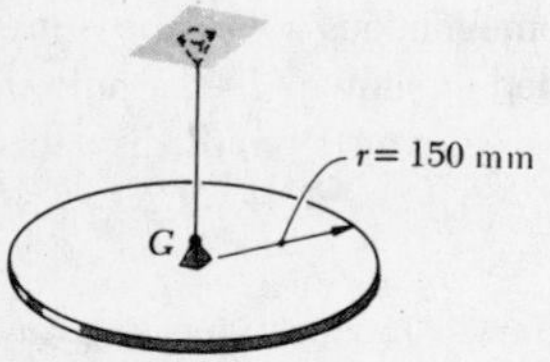

Fig. P19.29

19.30 A torsion pendulum may be used to determine experimentally the moment of inertia of a given object. The horizontal platform P is held by several rigid bars which are connected to a vertical wire. The period of oscillation of the platform is found equal to τ_0 when the platform is empty and to τ_A when an object of known moment of inertia I_A is placed on the platform so that its mass center is directly above the center of the plate. (*a*) Show that the moment of inertia I_0 of the platform and its supports may be expressed as $I_0 = I_A\tau_0^2/(\tau_A^2 - \tau_0^2)$. (*b*) If a period of oscillation τ_B is observed when an object B of unknown moment of inertia I_B is placed on the platform, show that $I_B = I_A(\tau_B^2 - \tau_0^2)/(\tau_A^2 - \tau_0^2)$.

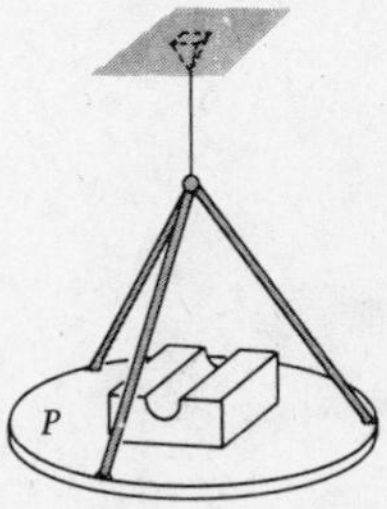

Fig. P19.30

19.31 A slender rod of weight 6 lb is suspended from a steel wire which is known to have a torsional spring constant $K = 0.25$ lb · in./rad. If the rod is rotated through 180° about the vertical and then released, determine (*a*) the period of oscillation, (*b*) the maximum velocity of end A of the rod.

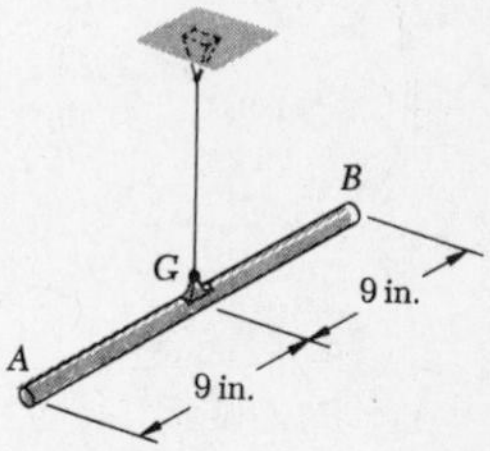

Fig. P19.31

19.32 A period of 4.10 s is observed for the angular oscillations of a 1-lb gyroscope rotor suspended from a wire as shown. Knowing that a period of 6.20 s is obtained when a 2-in.-diameter steel sphere is suspended in the same fashion, determine the centroidal radius of gyration of the rotor. (Specific weight of steel = 490 lb/ft³.)

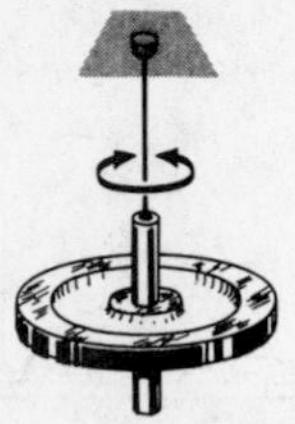
Fig. P19.32

19.33 Using the method of Sec. 19.6, solve Prob. 19.7.

19.34 Using the method of Sec. 19.6, solve Prob. 19.8.

19.35 Using the method of Sec. 19.6, solve Prob. 19.4.

19.36 Using the method of Sec. 19.6, solve Prob. 19.15.

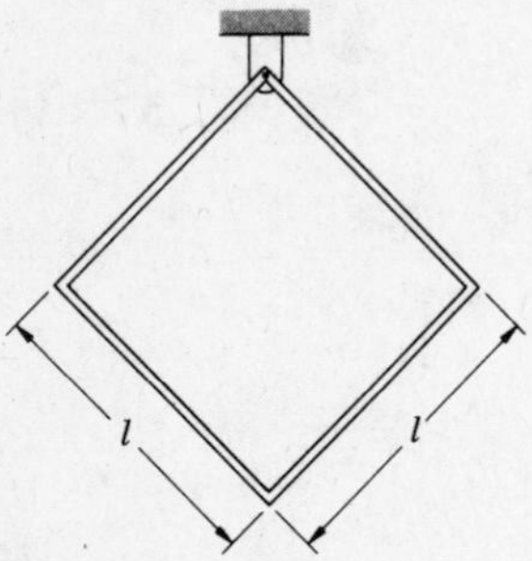

Fig. P19.37

19.37 A thin homogeneous wire is bent into the shape of a square of side l and suspended as shown. Determine the period of oscillation when the wire figure is given a small displacement to the right and released.

19.38 A thin homogeneous wire is bent into the shape of an equilateral triangle of side l. Determine the period of small oscillations if the wire figure is suspended as shown.

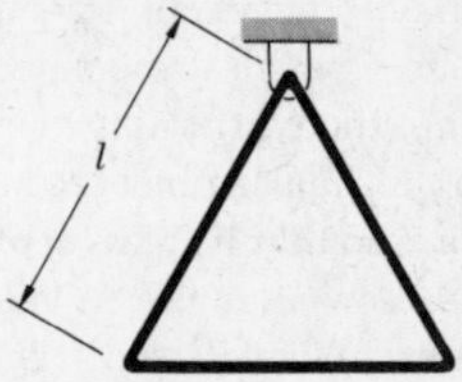

Fig. P19.38

19.39 Solve Prob. 19.38, assuming that the wire triangle is suspended from a pin located at the midpoint of one side.

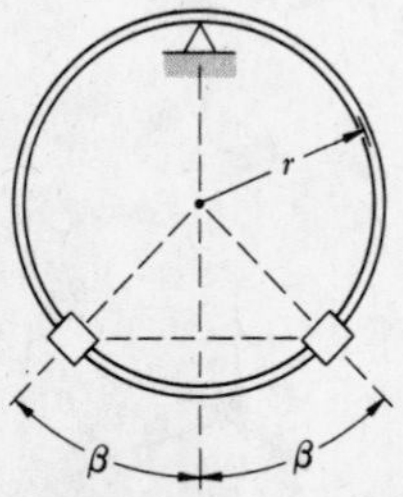

Fig. P19.40

19.40 Two collars, each of weight W, are attached as shown to a hoop of radius r and of negligible weight. (*a*) Show that for any value of β the period is $\tau = 2\pi\sqrt{2r/g}$. (*b*) Show that the same result is obtained if the weight of the hoop is not neglected.

19.41 The motion of the uniform rod AB is guided by the cord BC and by the small roller at A. Determine the frequency of oscillation when the end B of the rod is given a small horizontal displacement and released.

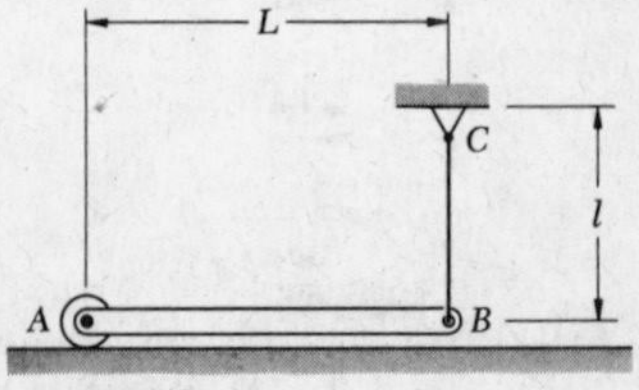

Fig. P19.41

19.42 A uniform disk of weight W and radius r may roll on the horizontal surface shown. In case a, a weight W of negligible dimensions is bolted to the disk at a distance $\frac{1}{2}r$ from the geometric center of the disk. In case b, a vertical force of constant magnitude W is applied to a cord which is attached to the disk at a distance $\frac{1}{2}r$ from the center of the disk. Determine the ratio τ_b/τ_a of the two periods of oscillation.

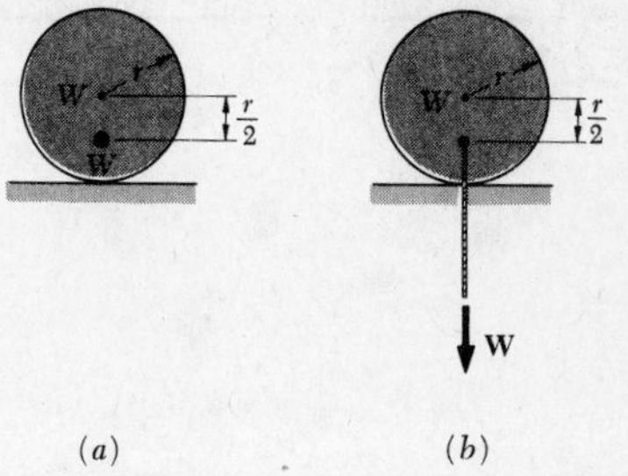

Fig. P19.42

19.43 Blade AB of the experimental wind-turbine generator shown is to be temporarily removed. Motion of the turbine generator about the y axis is prevented, but the remaining three blades may oscillate as a unit about the x axis. Assuming that each blade is equivalent to a 40-ft slender rod, determine the period of small oscillations of the blades.

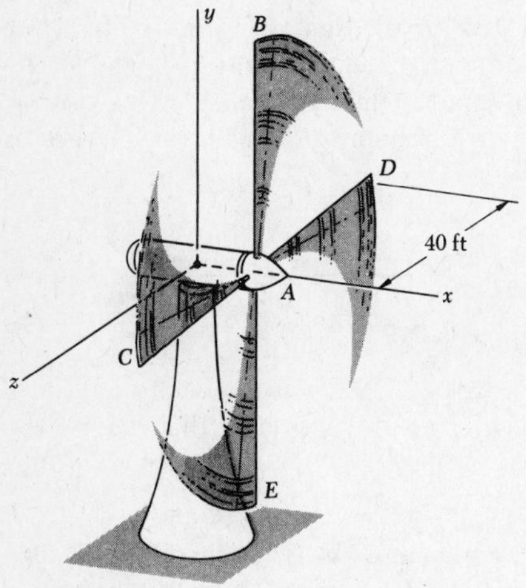

Fig. P19.43

19.44 Using the method of Sec. 19.6, solve Prob. 19.21.

19.45 Using the method of Sec. 19.6, solve Prob. 19.22.

19.46 The slender rod AB of mass m is attached to two collars of negligible mass. Knowing that the system lies in a horizontal plane and is in equilibrium in the position shown, determine the period of vibration if the collar A is given a small displacement and released.

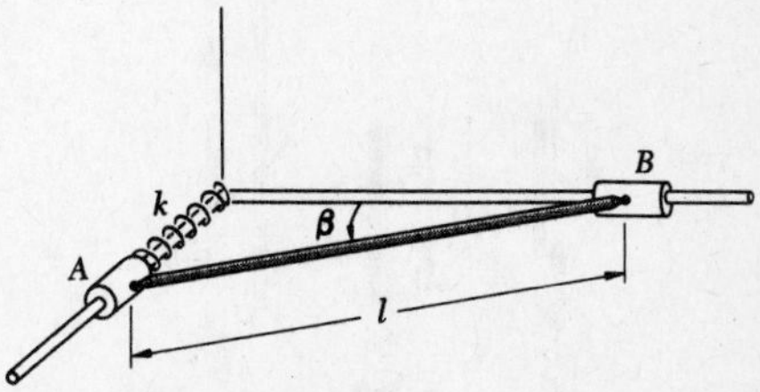

Fig. P19.46 and P19.48

19.47 Solve Prob. 19.46, assuming that rod AB is of mass m and that each collar is of mass m_c.

19.48 A slender rod AB of negligible mass connects two collars, each of mass m_c. Knowing that the system lies in a horizontal plane and is in equilibrium in the position shown, determine the period of vibration if the collar at A is given a small displacement and released.

SECTION 19.7

Fig. P19.49

19.49 A motor of mass 45 kg is supported by four springs, each of constant 100 kN/m. The motor is constrained to move vertically, and the amplitude of its movement is observed to be 0.5 mm at a speed of 1200 rpm. Knowing that the mass of the rotor is 14 kg, determine the distance between the mass center of the rotor and the axis of the shaft.

19.50 In Prob. 19.49, determine the amplitude of the vertical movement of the motor at a speed of (*a*) 200 rpm, (*b*) 1600 rpm, (*c*) 900 rpm.

19.51 A 500-lb motor is supported by a light horizontal beam. The unbalance of the rotor is equivalent to a weight of 1 oz located 10 in. from the axis of rotation. Knowing that the static deflection of the beam due to the weight of the motor is 0.220 in., determine (*a*) the speed (in rpm) at which resonance will occur, (*b*) the amplitude of the steady-state vibration of the motor at a speed of 800 rpm.

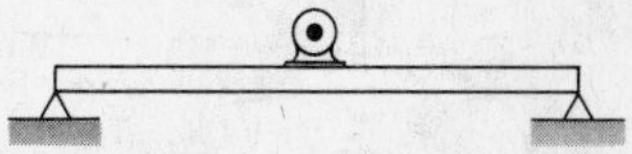

Fig. P19.51

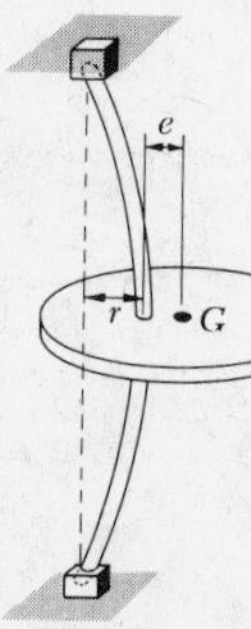

Fig. P19.53

19.52 Solve Prob. 19.51, assuming that the 500-lb motor is supported by a nest of springs having a total constant of 400 lb/in.

19.53 A disk of mass 30 kg is attached to the midpoint of a shaft. Knowing that a static force of 200 N will deflect the shaft 0.6 mm, determine the speed of the shaft in rpm at which resonance will occur.

19.54 Knowing that the disk of Prob. 19.53 is attached to the shaft with an eccentricity $e = 0.2$ mm, determine the deflection r of the shaft at a speed of 900 rpm.

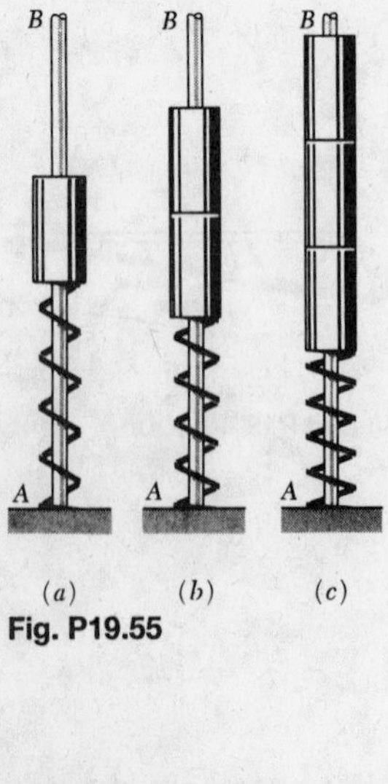

Fig. P19.55

19.55 Rod *AB* is rigidly attached to the frame of a motor running at a constant speed. When a collar of mass m is placed on the spring, it is observed to vibrate with an amplitude of 0.5 in. When two collars, each of mass m, are placed on the spring, the amplitude is observed to be 0.6 in. What amplitude of vibration should be expected when three collars, each of mass m, are placed on the spring? (Obtain two answers.)

19.56 Solve Prob. 19.55, assuming that the speed of the motor is changed and that one collar has an amplitude of 0.60 in. and two collars have an amplitude of 0.20 in.

19.57 A variable-speed motor is rigidly attached to the beam BC. When the speed of the motor is less than 1000 rpm or more than 2000 rpm, a small object placed at A is observed to remain in contact with the beam. For speeds between 1000 and 2000 rpm the object is observed to "dance" and actually to lose contact with the beam. Determine the speed at which resonance will occur.

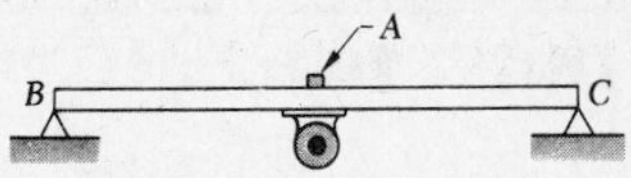

Fig. P19.57

19.58 As the speed of a spring-supported motor is slowly increased from 150 to 200 rpm, the amplitude of the vibration due to the unbalance of the rotor is observed to decrease continuously from 0.150 to 0.080 in. Determine the speed at which resonance will occur.

19.59 In Prob. 19.58, determine the speed for which the amplitude of the vibration is 0.200 in.

19.60 The amplitude of the motion of the pendulum bob shown is observed to be 3 in. when the amplitude of the motion of collar C is $\frac{3}{4}$ in. Knowing that the length of the pendulum is $l = 36$ in., determine the two possible values of the frequency of the horizontal movement of the collar C.

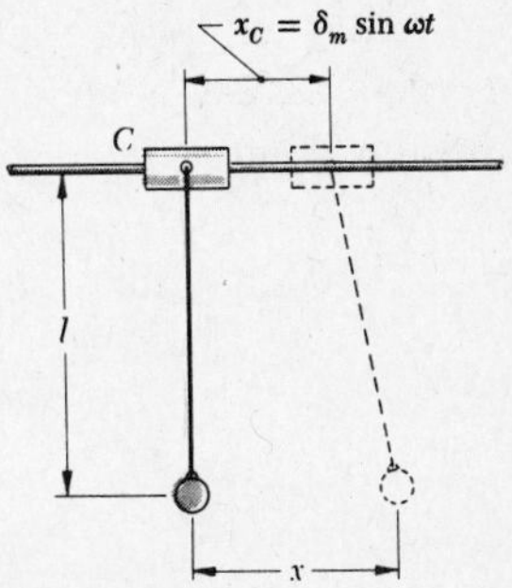

Fig. P19.60

19.61 A certain vibrometer used to measure vibration amplitudes consists essentially of a box containing a slender rod to which a mass m is attached; the natural frequency of the mass-rod system is known to be 5 Hz. When the box is rigidly attached to the casing of a motor rotating at 600 rpm, the mass is observed to vibrate with an amplitude of 1.6 mm relative to the box. Determine the amplitude of the vertical motion of the motor.

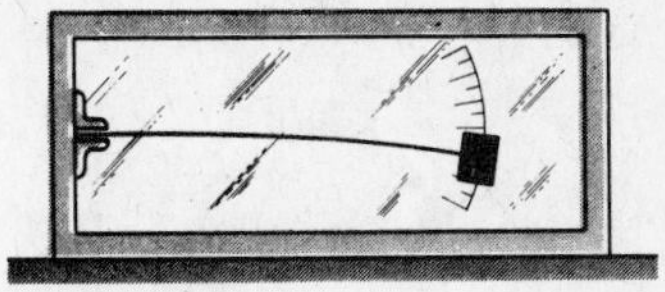
Fig. P19.61

19.62 A small trailer of mass 200 kg with its load is supported by two springs, each of constant 20 kN/m. The trailer is pulled over a road, the surface of which may be approximated by a sine curve of amplitude 30 mm and of period 5 m (i.e., the distance between two successive crests is 5 m, and the vertical distance from a crest to a trough is 60 mm). Determine (*a*) the speed at which resonance will occur, (*b*) the amplitude of the vibration of the trailer at a speed of 60 km/h.

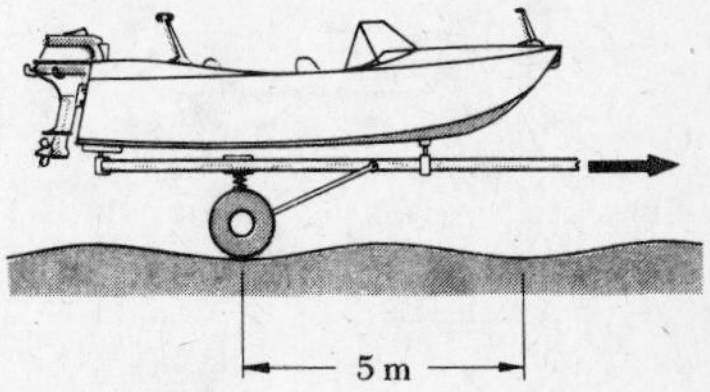

Fig. P19.62

19.63 Knowing that the amplitude of the vibration of the trailer of Prob. 19.62 is not to exceed 15 mm, determine the smallest speed at which the trailer can be pulled over the road.

19.64 A 2-kg instrument is spring-mounted on the casing of a motor rotating at 1800 rpm. The motor is unbalanced and the amplitude of the motion of its casing is 0.5 mm. Knowing that $k = 9000$ N/m, determine the amplitude of the motion of the instrument.

SECTIONS 19.8 to 19.10

19.101 Successive maximum displacements of a spring-mass-dashpot system similar to that shown in Fig. 19.10 are 75, 60, 48, and 38.4 mm. Knowing that $m = 20$ kg and $k = 800$ N/m, determine (*a*) the damping factor c/c_c, (*b*) the value of the coefficient of viscous damping c. (*Hint:* See Probs. 19.110 and 19.111 in main text.)

19.66 The barrel of a field gun weighs 1,400 lb and is returned into firing position after recoil by a recuperator of constant $k = 10{,}000$ lb/ft. Determine the value of the coefficient of damping of the recoil mechanism which causes the barrel to return into firing position in the shortest possible time without oscillation.

19.67 A critically damped system is released from rest at an arbitrary position x_0 when $t = 0$. (*a*) Determine the position of the system at any time t. (*b*) Apply the result obtained in part *a* to the barrel of the gun of Prob. 19.66, and determine the time at which the barrel is halfway back to its firing position.

19.68 Assuming that the barrel of the gun of Prob. 19.66 is modified, with a resulting increase in weight of 400 lb, determine the constant k of the recuperator which should be used if the recoil mechanism is to remain critically damped.

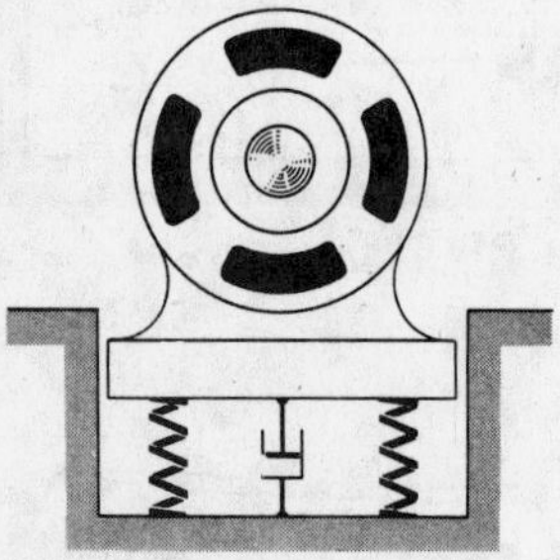

Fig. P19.69

19.69 A motor of mass 25 kg is supported by four springs, each having a constant of 200 kN/m. The unbalance of the rotor is equivalent to a mass of 30 g located 125 mm from the axis of rotation. Knowing that the motor is constrained to move vertically, determine the amplitude of the steady-state vibration of the motor at a speed of 1800 rpm, assuming (*a*) that no damping is present, (*b*) that the damping factor c/c_c is equal to 0.125.

19.70 Assume that the 25-kg motor of Prob. 19.69 is directly supported by a light horizontal beam. The static deflection of the beam due to the weight of the motor is observed to be 5.75 mm, and the amplitude of the vibration of the motor is 0.5 mm at a speed of 400 rpm. Determine (*a*) the damping factor c/c_c, (*b*) the coefficient of damping c.

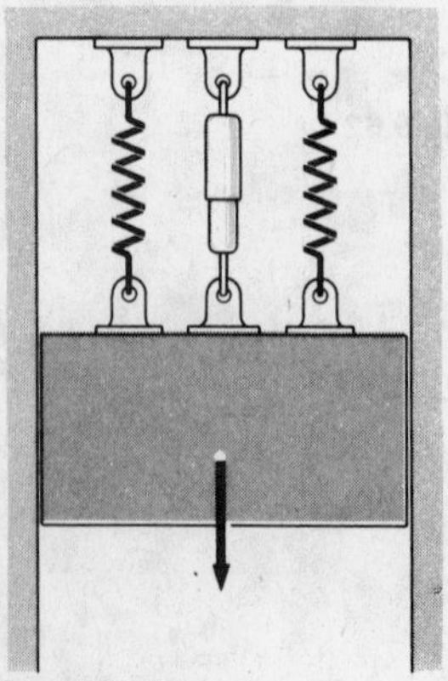

Fig. P19.71

19.71 A machine element weighing 800 lb is supported by two springs, each having a constant of 200 lb/in. A periodic force of maximum value 30 lb is applied to the element with a frequency of 2.5 cycles per second. Knowing that the coefficient of damping is 8 lb · s/in., determine the amplitude of the steady-state vibration of the element.

19.72 In Prob. 19.71, determine the required value of the coefficient of damping if the amplitude of the steady-state vibration of the element is to be 0.15 in.

19.73 A platform of mass 100 kg, supported by a set of springs equivalent to a single spring of constant $k = 80$ kN/m, is subjected to a periodic force of maximum magnitude 500 N. Knowing that the coefficient of damping is 2 kN · s/m, determine (*a*) the natural frequency in rpm of the platform *if* there were no damping, (*b*) the frequency in rpm of the periodic force corresponding to the maximum value of the magnification factor, assuming damping, (*c*) the amplitude of the *actual* motion of the platform for each of the frequencies found in parts *a* and *b*.

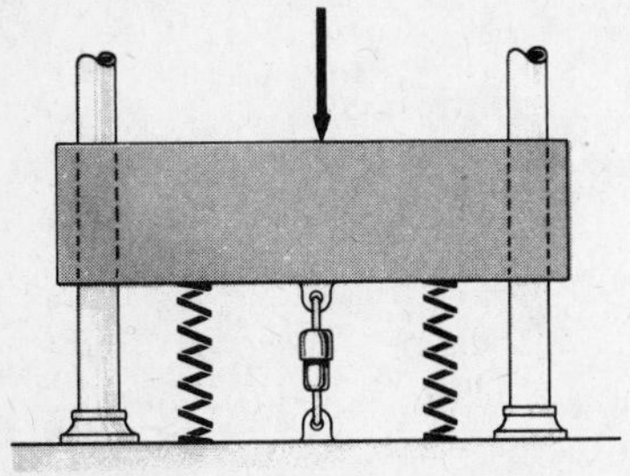

Fig. P19.73

***19.74** Two loads *A* and *B*, each of mass *m*, are suspended as shown by means of five springs of the same constant *k*. Load *B* is subjected to a force of magnitude $P = P_m \sin \omega t$. Write the differential equations defining the displacements x_A and x_B of the two loads from their equilibrium positions.

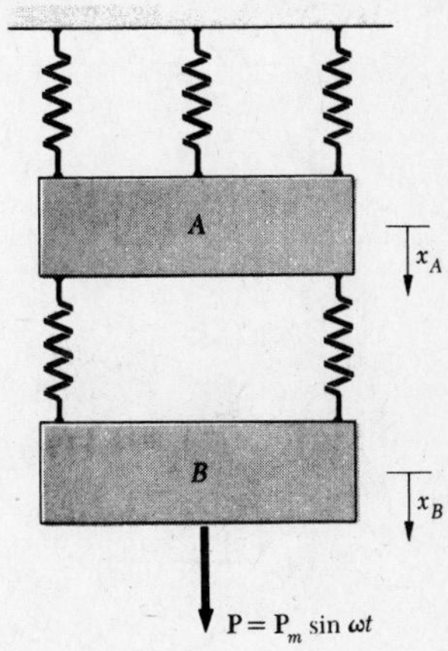

Fig. P19.74

19.75 and 19.76 Draw the electrical analogue of the mechanical system shown.

19.77 and 19.78 Write the differential equations defining (*a*) the displacements of the masses m_1 and m_2, (*b*) the currents in the corresponding loops of the electrical analogue.

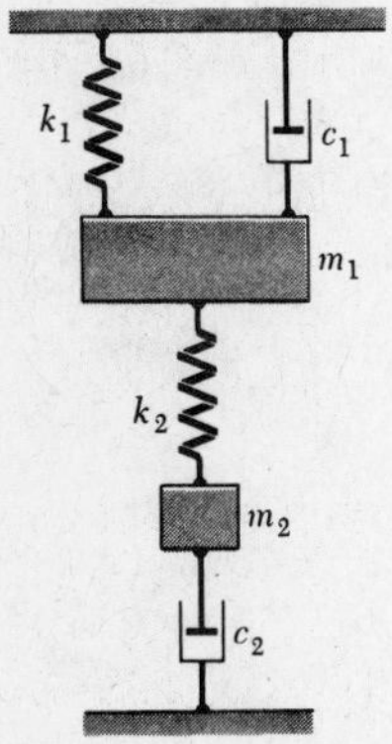

Fig. P19.75 and P19.77

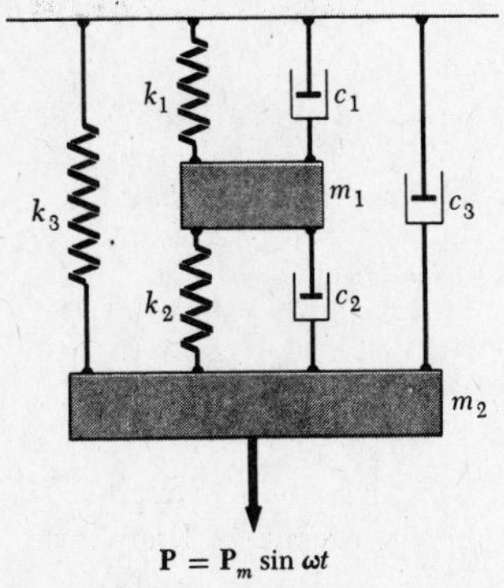

Fig. P19.76 and P19.78

ANSWERS TO PROBLEMS

CHAPTER 11

11.1 $t = 0$, $x = 12$ in., $a = -18$ in./s^2; $t = 3$ s, $x = -15$ in., $a = 18$ in./s^2.

11.2 $s = 22$ in., $v = 12$ in./s, $a = 18$ in./s^2.

11.3 (*a*) 5 s. (*b*) 14 m, 34 m.

11.4 (*a*) 2 s, 4 s. (*b*) 8 m, 7.33 m.

11.5 $a = 2t$, $v = t^2 - 16$, $x = \frac{1}{3}t^3 - 16t + 42.7$.

11.6 -4 m/s; 12 m; 20 m.

11.7 (*a*) 12 in./s^4. (*b*) $a = 12t^2$, $v = 4t^3 - 250$, $x = t^4 - 250t + 484$.

11.8 (*a*) 3 s. (*b*) 116 in., -56 in./s. (*c*) 65 in.

11.9 (*a*) ± 13.35 in./s. (*b*) 30.35 in.

11.10 (*a*) 384 in^3/s^2. (*b*) 13.86 in./s.

11.11 (*a*) ± 6 m/s. (*b*) $+2.29$ m. (*c*) 1.323 m.

11.12 25 s^{-2}.

11.13 (*a*) 3 m. (*b*) Infinite. (*c*) 0.461 s.

11.14 (*a*) 55.5 m. (*b*) Infinite.

11.15 142.7 ft/s.

11.16 (*a*) 173.3 ft. (*b*) Infinite.

11.17 (*a*) 5 m/s. (*b*) 11 m/s. (*c*) 60 m.

11.18 (*a*) 29.1 m. (*b*) 24.4 m/s.

11.19 (*a*) 30.7 ft/s. (*b*) 98.2 ft/s.

11.20 (*a*) -1.689 ft/s^2. (*b*) 40.7 ft/s.

11.21 (*a*) 60 s; 960 m. (*b*) 240 s; 5280 m.

11.22 (*a*) 1.115 s; 35.9 ft/s $\downarrow$. (*b*) 2.74 s; 14.57 ft/s $\downarrow$. (*c*) 59.1.

11.23 (*a*) 17.10 s; 171.0 m. (*b*) 81.5 km/h.

11.24 (*a*) 1.200 m.

11.25 $t = 15$ s; $x = 450$ ft.

11.26 (*b*) 0.994 s; 15.90 ft.

11.27 (*a*) 12 in./s $\downarrow$, 12 in./s $\uparrow$, 36 in./s $\downarrow$, 36 in./s $\uparrow$. (*b*) 48 in./s $\uparrow$. (*c*) 48 in./s $\uparrow$.

11.28 (*a*) 1.5 in./s $\downarrow$, 4.5 in./s $\uparrow$. (*b*) 6 in./s $\downarrow$, 12 in. $\downarrow$.

11.29 (*a*) 120 mm/s $\uparrow$. (*b*) 120 mm/s $\downarrow$.

11.30 (*a*) $a_A = 45$ mm/s$^2 \rightarrow$; $a_B = 30$ mm/s$^2 \rightarrow$. (*b*) $v_B = 90$ mm/s$\rightarrow$; $x_B = 135$ mm$\rightarrow$.

11.31 (*a*) 3 s. (*b*) 84.4 mm $\uparrow$.

11.32 (*a*) $v_C = \frac{3}{4}\, v_A + \frac{1}{4}\, v_B$ (*b*) 375 mm/s $\uparrow$, 750 mm $\uparrow$.

11.33 $v_A = 3$ ft/s $\downarrow$; $v_B = 5$ ft/s $\uparrow$; $v_C = 4$ ft/s $\downarrow$.

11.34 (*a*) 0.500 s. (*b*) $\frac{1}{16}$ ft $\uparrow$.

11.35 (*a*) 72 m. (*b*) 4 s; 15 s.

11.36 (*a*) 48 m. (*b*) 6 s, 13.75 s, 16.25 s.

11.37 (*b*) 10 ft/s; 182 ft; 214 ft.

11.38 (*a*) 32 ft/s. (*b*) 192 ft.

11.39 (*a*) 42 s. (*b*) 1848 ft.

11.40 (*a*) 7.60 ft/s^2. (*b*) 8.18 mi/h.

11.41 (*a*) 60 s; 960 m. (*b*) 240 s; 5280 m.

11.42 (*a*) Local. (*b*) 122 s.

11.43 (*a*) 16 s. (*b*) 310 m.

11.44 (*a*) 8.57 s. (*b*) 1.867 m/s^2; 1.400 m/s^2. (*c*) 68.6 m; 51.4 m.

11.45 (*a*) 2 s. (*b*) 1500 ft.

11.46 25 s.

11.47 4.28 m/s; 0.188 m.

11.48 (*a*) 10.0 m/s; 27.4 m. (*b*) 13.9 m/s; 51.5 m.

11.49 (*a*) 130 ft/s. (*b*) 316 ft.

11.50 (*a*) 122 ft/s. (*b*) 296 ft.

11.51 (*a*) $\mathbf{v} = 2.69$ m/s $\angle 21.8°$;
$\mathbf{a} = 1.414$ m/s^2 $\angle 45°$.
(*b*) $\mathbf{v} = 1.803$ m/s $\angle 33.7°$;
$\mathbf{a} = 5.10$ m/s^2 $\angle 11.3°$.

11.52 (*a*) 2 s. (*b*) 2 m/s$\leftarrow$; 2.24 m/s^2 $\angle 26.6°$.

11.53 (*a*) $\mathbf{v} = 0.707$ ft/s $\angle 45°$;
$\mathbf{a} = 0.354$ ft/s^2 $\angle 45°$.
(*b*) $\mathbf{v} = 0.878$ ft/s $\angle 20.2°$;
$\mathbf{a} = 0.439$ ft/s^2 $\angle 20.2°$.

11.54 $\mathbf{v} = 2.22$ ft/s $\angle 34.2°$;
$\mathbf{a} = 2.22$ ft/s^2 $\angle 34.2°$.

11.55 4.20 m/s $\leqslant v_0 \leqslant$ 6.64 m/s.

11.56 (*a*) 4.79 m/s $\angle 50°$. (*b*) 0.975 s.

11.57 12.43 ft

11.58 15.8° or −54.5°.

11.59 15° or 75°.

11.60 1609 ft.

11.61 18.54 mi/h from 62.8° west of south.

11.62 23.2 miles, 62.8° west of south.

11.63 (*a*) 187.4 km/h $\angle 63.7°$.
(*b*) 208 m $\angle 63.7°$. (*c*) 353 m.

11.64 (*a*) 56.3° from rear of truck.
(*b*) 16.63 m/s.

11.65 (*a*) 63.9°. (*b*) 1195 ft/s $\uparrow$, 32.2 ft/s^2 $\downarrow$.

11.66 (*a*) 21.1°. (*b*) 1.02°.

11.67 623 m/s $\angle 36.7°$; 9.81 m/s^2 $\downarrow$.

11.68 10.18 m/s $\angle 10.8°$; 9.81 m/s^2 $\downarrow$.

11.69 103.9 m.

11.70 1.2 m/s^2.

11.71 1.926 ft/s^2.

11.72 8.51 ft/s^2.

11.73 (*a*) 2009 ft. (*b*) 1811 ft.

11.74 (*a*) 79.6 m. (*b*) 40.8 m.

11.75 22,800 ft; 58,100 ft.

11.76 (*a*) 111.8 ft. (*b*) 124.1 ft.

11.77 2585 km.

11.78 27 400 km/h.

11.79 2280 mi/h.

11.80 (*a*) $\mathbf{v} = 0$; $\mathbf{a} = (120\text{ mm/s}^2)\mathbf{e}_r$.
(*b*) $\mathbf{v} = (60\text{ mm/s})\mathbf{e}_r + (160\text{ mm/s})\mathbf{e}_\theta$;
$\mathbf{a} = -(640\text{ mm/s}^2)\mathbf{e}_r + (640\text{ mm/s}^2)\mathbf{e}_\theta$.

11.81 $\mathbf{v} = -(180\text{ mm/s})\mathbf{e}_r$;
$\mathbf{a} = -(0.240\text{ m/s}^2)\mathbf{e}_r - (4.32\text{ m/s}^2)\mathbf{e}_\theta$.

11.82 (*a*) $\mathbf{v} = -4\pi b\mathbf{e}_r + 4\pi b\mathbf{e}_\theta$;
$\mathbf{a} = -8\pi^2 b\mathbf{e}_r - 16\pi^2 b\mathbf{e}_\theta$.
(*b*) $\mathbf{v} = 0$; $\mathbf{a} = 8\pi^2 b\mathbf{e}_r$.

11.83 (*a*) $\mathbf{v} = (\frac{1}{2}\pi b)\mathbf{e}_\theta$; $\mathbf{a} = -(\frac{5}{4}\pi^2 b)\mathbf{e}_r$.
(*b*) $\mathbf{v} = -(\pi b)\mathbf{e}_r$; $\mathbf{a} = -(\pi^2 b)\mathbf{e}_\theta$.

11.84 (*a*) $\dot\theta = 2.94v_0$; $v_r = 0$.
(*b*) $\dot\theta = 1.479v_0$; $v_r = -12v_0/13$.

CHAPTER 12

12.1 32.6 N.

12.2 (*a*) 3.37 m/s. (*b*) 10.28 m.

12.3 $F_{AB} = 500$ lb C; $F_{BC} = 500$ lb T.

12.4 (*a*) $\mathbf{a}_A = \mathbf{a}_B = 2.42$ ft/s$^2\swarrow$. (*b*) 1.160 lb$\nearrow$.

12.5 3.39 m/s^2 $\angle 60°$.

12.6 (*a*) 302 N. (*b*) 6.79 m/s $\uparrow$.
(*c*) 1.346 m/s $\downarrow$.

12.7 (*a*) 50.2 kg. (*b*) 357 N.

12.8 (*a*) 24.9 lb$\rightarrow$. (*b*) 7.96 lb.

12.9 0.566 lb$\leftarrow$.

12.10 (*a*) 10.73 ft/s$^2\rightarrow$. (*b*) 6.33 lb.

12.11 (*a*) 122.6 N. (*b*) 2.45 m/s$^2\leftarrow$.

12.12 (*a*) 4.56 m/s$^2\leftarrow$. (*b*) 1.962 m/s$^2\leftarrow$.
(*c*) 2.60 m/s$^2\rightarrow$.

12.13 (*a*) and (*b*) 9.66 ft/s$^2\leftarrow$.

12.14 $\mathbf{a}_A = 4.41$ m/s^2 $\downarrow$; $\mathbf{a}_B = 0.981$ m/s^2 $\uparrow$;
$\mathbf{a}_C = 2.45$ m/s$^2\leftarrow$.

12.15 $\mathbf{a}_A = 4.91$ m/s^2 $\uparrow$; $\mathbf{a}_B = 2.45$ m/s^2 $\downarrow$;
$\mathbf{a}_C = 0$.

12.16 $W_B = 2.60\ W_A$; $W_C = 1.342\ W_A$.

12.17 $\mathbf{a}_A = 13.26$ ft/s^2 $\uparrow$; $\mathbf{a}_B = 1.894$ ft/s^2 $\downarrow$;
$\mathbf{a}_C = 9.47$ ft/s^2 $\downarrow$. Block C strikes ground first.

12.18 1.656 m/s; 9.16 m/s^2.

12.19 8.02 ft/s $< v <$ 13.90 ft/s.

12.20 (*a*) 10.56 ft/s. (*b*) 7.32 lb.

12.21 (*a*) 64.0 N. (*b*) 3.25 m/s.

12.22 A: 12.86 ft/s^2. B: 25.8 ft/s^2. C: 19.32 ft/s^2.

12.23 2.71 m/s.

12.24 $-62.7° < \theta < 62.7°$.

12.25 (*a*) W. (*b*) $\frac{1}{2}W$; 0.866g $\angle 60°$.

12.26 $v_0 = \sqrt{gy}$.

12.27 0.815.

12.28 22.5°.

12.29 (*a*) 1790 N. (*b*) Impossible.

12.30 (*a*) 3.11 m/s^2 $\downarrow$. (*b*) 9.81 m/s^2 $\downarrow$.

12.31 (*a*) $\mathbf{a}_A = 8.92$ ft/s$^2\leftarrow$; $\mathbf{a}_B = 5.94$ ft/s$^2\leftarrow$.
(*b*) 3.08 lb.

12.32 (*a*) 64.2 lb. (*b*) Impossible, since a_B cannot be greater than g.

12.33 (*a*) $F_r = 4$ N, $F_\theta = 0$.
(*b*) $F_r = -21.3$ N, $F_\theta = 21.3$ N.

12.34 $F_r = -8$ N; $F_\theta = -144$ N.
12.35 (*a*) $F_r = -2.30$ lb, $F_\theta = 0$.
(*b*) $F_r = 0$, $F_\theta = -1.839$ lb.
12.36 $n = 0$: uniform circular motion;
$n = 1$: uniform rectilinear motion.
12.37 (*a*) 24 in./s. (*b*) $\rho_A = \frac{2}{3}$ in., $\rho_B = 18$ in.
12.38 409×10^{21} lb · s²/ft or 13.17×10^{24} lb.
12.39 111.4 min.
12.40 (*a*) 1.804×10^{-6} N. (*b*) 191 s.
12.41 (*a*) 5880 km/h. (*b*) 5630 km/h.
12.42 140 mm.
12.43 2640 mi/h.
12.44 (*a*) 6.95 in. (*b*) 17.27 ft/s.
12.45 1003 m/s.
12.46 −30.4 m/s.
12.47 12,160 mi.
12.48 4630 mi.
12.49 101 m/s.
12.50 32 800 km/h; 14 770 km/h.
12.51 (*a*) 5590 ft/s. (*b*) 94 ft/s.
12.52 (*a*) 5560 ft/s. (*b*) 61 ft/s.
12.53 46 h 33 min.
12.54 3 h 2 min.
12.55 57 min 35 s.
12.56 116.6 h.
12.57 79.7°.
12.58 4560 m/s.
12.59 197 ft/s.
12.60 Angle $BOC = 44.8°$.

CHAPTER 13

13.1 16.98 ft/s.
13.2 19.99 N.
13.3 14.40 N.
13.4 321 lb.
13.5 12.67 ft/s←.
13.6 5.17 ft/s←.
13.7 5.29 m/s→.
13.8 45.9 N→.
13.9 1.981 m/s.
13.10 3.71 m/s.
13.11 (*a*) 9.27 ft/s↑. (*b*) 9.33 ft.
13.12 (*a*) 13.10 ft/s↑. (*b*) 10.67 ft.
13.13 0.485.
13.14 (*a*) 10.91 hp. (*b*) 50.9 hp.
13.15 (*a*) 6.87 kW. (*b*) 8.44 kW.
13.16 0.076 m/s² ↑.
13.17 46.6 ft/s.
13.18 9.66 lb.
13.19 2.45 m/s.
13.20 (*a*) 4.71 m/s. (*b*) 4.03 m/s.
13.21 180.5 N.
13.22 104.9 N↓.
13.23 (*a*) 12.55 ft/s.
(*b*) $\mathbf{R} = -(4 \text{ lb})\mathbf{i} + (12.5 \text{ lb})\mathbf{j}$.
13.24 3.21 lb.
13.25 (*a*) 4.91 ft/s. (*b*) 0.250 lb.
13.26 (*a*) 22.7 ft/s. (*b*) 7.75 ft.
13.27 (*b*) $b = \frac{1}{2}h$; $c_{max} = h$.
13.28 6 *mg*.
13.29 317 N/m.
13.30 233 N/m.
13.32 $V = \ln\sqrt{x^2 + y^2}$.
13.33 (*a*) $2a^2$. (*b*) $3a^2/2$. (*c*) a^2.
13.34 Paths *ABC* and *ADC*, $U = a^4/3$;
path *AC*, $U = a^4/2$.
13.35 (*a*) 33.5×10^6 J/kg. (*b*) 45.0×10^6 J/kg.
13.36 (*a*) 111.9×10^9 ft · lb.
(*b*) 149.7×10^9 ft · lb.
13.37 (*a*) $\frac{1}{2}mgR$. (*b*) $\frac{3}{4}mgR$.
13.38 (*a*) 4.50 m/s. (*b*) 1.500 m/s. (*c*) 37.5 mm.
13.39 (*a*) $v_2 = 2v_1$. (*b*) +100 per cent. (*c*) Zero.
(*d*) +300 per cent.
13.40 (*a*) $v_2 = v_1$. (*b*) Zero. (*c*) −50%. (*d*) Zero.
13.41 (*a*) 27.3 ft/s. (*b*) 1.099 ft.
13.42 2.84 ft/s.
13.46 50.9°.
13.47 8420 m/s; 74.4°.
13.48 2280 m/s.
13.49 11,280 miles, 1018 miles.
13.50 $65.6° \leq \phi_0 \leq 114.4°$.
13.51 (*a*) and (*b*) 6 min 4 s.
13.52 219 s.
13.53 (*a*) 2.65 s. (*b*) 21.2 s.
13.54 $\mathbf{v} = (120 \text{ m/s})\mathbf{i} + (300 \text{ m/s})\mathbf{j} + (285 \text{ m/s})\mathbf{k}$.
13.55 (*a*) 3.06 s. (*b*) 6.12 s.
13.56 (*a*) 2.80 s. (*b*) 5.60 s.
13.57 (*a*) 19.13 s. (*b*) 714 lb *C*.
13.58 (*a*) 38.3 s. (*b*) 2140 lb T.
13.59 227 N←, 818 N↓.
13.60 2.86 m/s.
13.61 (*a*) 4 ft/s←. (*b*) 49.7 lb←.
13.62 (*a*) 4.93 m/s. (*b*) 0.01408.
13.63 (*a*) 0.375 mi/h. (*b*) 4270 lb.

13.64 (*a*) 0.625 mi/h. (*b*) 4270 lb.
13.65 (*a*) 2 m/s←. (*b*) $T_A = 3$ J. $T_B = 9$ J.
13.66 (*a*) 24.9 lb · s. (*b*) 19.94 lb · s.
(*c*) 200 ft · lb; 160 ft · lb.
13.67 0.400 lb.
13.68 0.500.
13.69 (*a*) $\mathbf{v}_A = 0.563$ m/s←; $\mathbf{v}_B = 6.94$ m/s→.
(*b*) 41.0 J.
13.70 (*a*) $\mathbf{v}_A = 3.38$ m/s→; $\mathbf{v}_B = 6.38$ m/s→.
(*b*) 6.56 J.
13.71 (*a*) 0.571 $\mathbf{v}_0$. (*b*) 1.333 $\mathbf{v}_0$.
13.72 0.330.
13.73 $\mathbf{v}_A = 3.50$ m/s ∠60°;
$\mathbf{v}_B = 4.03$ m/s ∠21.7°.
13.75 $\tan^{-1}\sqrt{e}$.
13.76 (*a*) $\tan\theta = \sqrt{e}$. (*b*) $\sqrt{e}\, v_0$.
13.77 (*a*) $\cot\beta = \sqrt{e}$. (*b*) ev_0.
13.78 (*a*) 0.943. (*b*) 711 mm; 377 mm.
13.79 (*a*) 0.883. (*b*) 11.30 in.
13.80 69.9°.
13.81 (*a*) 76.2°. (*b*) 37.1 N.
13.82 (*a*) 8.29 ft/s→. (*b*) 6.85 lb. (*c*) 1.068 ft.
13.83 79.8°.
13.84 (*a*) 0.80. (*b*) 0.32.
13.85 (*a*) Five. (*b*) 2 m/s→. (*c*) Same as original.
13.86 (*a*) $x = l$. (*b*) $(1 + e)^4/16$.

CHAPTER 14

14.1 (*a*) 6 ft/s←. (*b*) 3 ft/s←; 7 ft/s←.
14.2 (*a*) 0. (*b*) 3 ft/s←; 1 ft/s→.
14.3 (*a*) 3.47 mi/h→. (*b*) 2.60 mi/h→.
14.4 (*a*) 2.88 mi/h→. (*b*) 2.60 mi/h→.
14.5 (*a*) $\frac{2}{3}v$; (*BC*) 33.3%. (*b*) $\frac{2}{3}v$; (*BC*) 25%, (*AB*) 8.33%.
14.6 $\mathbf{v}_A = 0.948$ m/s→; $\mathbf{v}_B = 1.092$ m/s→;
$\mathbf{v}_C = 1.960$ m/s→.
14.7 (*a*) $v_x = 19.50$ m/s, $v_y = 13.75$ m/s.
(*b*) $-(8.10\ \text{kg}\cdot\text{m}^2/\text{s})\mathbf{k}$.
14.8 (*a*) $v_x = 19.50$ m/s, $v_y = 16.00$ m/s.
(*b*) $-(5.40\ \text{kg}\cdot\text{m}^2/\text{s})\mathbf{i}$.
14.9 $x = 100$ m, $y = -40.7$ m, $z = 16$ m.
14.10 $\mathbf{r}_A = (200\ \text{ft})\mathbf{i} + (200\ \text{ft})\mathbf{j} + (1800\ \text{ft})\mathbf{k}$.
14.11 $\mathbf{v}_A = -(200\ \text{ft/s})\mathbf{j} + (600\ \text{ft/s})\mathbf{k}$.
14.12 (*a*) 5.15 lb. (*b*) 30.0 ft/s.
14.13 (*a*) 3.85 lb. (*b*) 376 ft/s.
14.14 $v_B = 1.143$ m/s; $v_C = 1.444$ m/s.
14.15 $-(32.75\ \text{m/s})\mathbf{j}$.
14.16 123.4 ft · lb.
14.17 9.57%.
14.18 3.08 kJ.
14.19 2440 ft · lb.
14.20 $\mathbf{r}_C = (96\ \text{ft})\mathbf{i} - (42\ \text{ft})\mathbf{j}$;
$\mathbf{v}_C = (36\ \text{ft/s})\mathbf{i} - (18\ \text{ft/s})\mathbf{j}$.
14.21 $v_A = v_0 \sin\theta$; $v_B = v_0 \sin\theta \cos\theta$;
$v_C = v_0 \cos^2\theta$.
14.22 $v_A = 1.500$ m/s; $v_B = 1.299$ m/s;
$v_C = 2.25$ m/s.
14.23 $-(100\ \text{m/s})\mathbf{j} + (300\ \text{m/s})\mathbf{k}$.
14.24 (*a*) $mv_0\mathbf{i}$; $\frac{3}{4}mlv_0\mathbf{k}$. (*b*) $\mathbf{v}_A = \frac{1}{4}v_0\mathbf{i} + \frac{3}{4}v_0\mathbf{j}$;
$\mathbf{v}_B = \frac{1}{4}v_0\mathbf{i} - \frac{1}{4}v_0\mathbf{j}$. (*c*) $\mathbf{v}_A = -\frac{1}{2}v_0\mathbf{i}$;
$\mathbf{v}_B = \frac{1}{2}v_0\mathbf{i}$.
14.25 $\mathbf{v}_A = 10.39$ m/s ∠30.0°;
$\mathbf{v}_B = 5.29$ m/s ∠40.9°.
14.26 (*a*) 8.67 ft/s→. (*b*) 4.81 ft/s ∠56.3°.
(*c*) $b = 5.83$ ft.
14.27 (*a*) 13.00 ft/s→. (*b*) 10.82 ft/s ∠33.7°.
(*c*) $b = 8.33$ ft.
14.28 (*a*) $\mathbf{v}_A = 3.20$ m/s ↑;
$\mathbf{v}_B = 5.30$ m/s ∠31.9°. (*b*) $b = 5.98$ m.
14.29 $P_x = 800$ N; $P_y = 800$ N.
14.30 $P_x = 170.6$ lb; $P_y = 170.6$ lb.
14.31 $\rho Q v$ ∠θ.
14.32 $Q_1 = \frac{1}{2}Q(1 - \sin\theta)$; $Q_2 = \frac{1}{2}Q(1 + \sin\theta)$.
14.33 $\mathbf{A}_x = A\rho v_A^2(1 - \cos\theta)$←, $\mathbf{A}_y =$
$A\rho v_A^2 \sin\theta$ ↑, $\mathbf{M}_A = A\rho v_A^2 R(1 - \cos\theta)$↺.
14.34 $\mathbf{C}_x = 475$ N←, $\mathbf{C}_y = 675$ N ↑;
$\mathbf{D} = 865$ N→.
14.35 $\mathbf{C}_x = 44.3$ lb←, $\mathbf{C}_y = 20$ lb ↑;
$\mathbf{D} = 44.3$ lb→.
14.36 $\mathbf{C}_x = 83.3$ lb←, $\mathbf{C}_y = 30.3$ lb ↓,
$\mathbf{M}_C = 496$ lb · in.↻.
14.37 (*a*) 32.4 m/s. (*b*) 101.9 m³/s. (*c*) 64.9 kJ/s.
14.38 (*a*) 26.4 kN. (*b*) 830 km/h.
14.39 6.31 lb.
14.40 35.2 ft/s.
14.41 (*a*) 24 rad/s↺. (*b*) 0.400 N · m↻.
14.42 (*a*) 1500 N. (*b*) 2500 N.
14.43 461 mi/h.
14.44 100.6 lb/s.
14.45 $\sin\theta = v^2/gL$.
14.46 (*a*) 28.2 N ↑. (*b*) 39.9 N ↑.
14.47 1553 lb.
14.49 (*a*) 53.7 lb/s. (*b*) 13.42 lb/s.

14.50 (a) 230 ft/s^2. (b) 1018 ft/s^2.
14.51 (a) $a = 25.2$ m/s^2; $v = 0$.
(b) $a = 53.8$ m/s^2; $v = 1651$ m/s.
(c) $a = 340$ m/s^2; $v = 7180$ m/s.
14.52 1.538 Mg/s.
14.53 18,480 mi/h.
14.54 $v = v_0 + u \ln \dfrac{m_0}{m_0 - qt} - gt$
14.55 (a) 6820 kg. (b) 341 s.
14.56 452,000 ft.
14.57 $h = u\left(t - \dfrac{m_0 - qt}{q} \ln \dfrac{m_0}{m_0 - qt}\right) - \dfrac{1}{2}gt^2$.
14.58 $e^{v_0/u} - 1$.

CHAPTER 15

15.1 (a) 10 rad; 4 rad/s; −4 rad/s^2.
(b) 7 rad; 11 rad/s; 14 rad/s^2.
15.2 (a) 2 s. (b) 2 rad; 8 rad/s^2.
15.3 (a) −2.51 rad/s^2. (b) 18 000 rev.
15.4 (a) −2.42 rad/s^2. (b) 52 s.
15.5 $\mathbf{v}_D = (0.14\text{ m/s})\mathbf{i} - (0.48\text{ m/s})\mathbf{j} - (0.624\text{ m/s})\mathbf{k}$; $\mathbf{a}_D = -(0.874\text{ m/s}^2)\mathbf{i} + (3.00\text{ m/s}^2)\mathbf{j} - (2.50\text{ m/s}^2)\mathbf{k}$.
15.6 $\mathbf{v}_E = (0.14\text{ m/s})\mathbf{i} - (0.48\text{ m/s})\mathbf{j} - (0.96\text{ m/s})\mathbf{k}$; $\mathbf{a}_E = -(0.644\text{ m/s}^2)\mathbf{i} + (2.21\text{ m/s}^2)\mathbf{j} - (7.30\text{ m/s}^2)\mathbf{k}$.
15.7 $\mathbf{v}_C = -(6\text{ in./s})\mathbf{i} + (12\text{ in./s})\mathbf{k}$; $\mathbf{a}_C = -(24\text{ in./s}^2)\mathbf{i} - (30\text{ in./s}^2)\mathbf{j} - (12\text{ in./s}^2)\mathbf{k}$.
15.8 $\mathbf{v}_C = -(6\text{ in./s})\mathbf{i} + (12\text{ in./s})\mathbf{k}$; $\mathbf{a}_C = -(12\text{ in./s}^2)\mathbf{i} - (30\text{ in./s}^2)\mathbf{j} - (36\text{ in./s}^2)\mathbf{k}$.
15.9 (a) 0.698 rad/s^2. (b) 15.00 s.
15.10 (a) 10 rad/s. (b) $\mathbf{a}_B = 18$ m/s^2 ↓; $\mathbf{a}_C = 6$ m/s^2 ↓.
15.11 (a) 5 rad/s^2↻; 2 rad/s↺. (b) 19.21 $in./s^2$.
15.12 (a) 6.93 rad/s↺. (b) 2.09 s.
15.13 (a) 19.10 rev. (b) 12.05 s.
15.14 (a) 24 rad/s^2↻. (b) 30.6 rev.
15.15 3.49 s; 6.98 s; 13.96 s.
15.16 (a) 3.95rad/s^2↺. (b) 3.18 s.
15.17 (a) 8.92 s. (b) $\boldsymbol{\omega}_A = 26.8$ rad/s↺; $\boldsymbol{\omega}_B = 44.6$ rad/s↻.
15.18 $\boldsymbol{\alpha}_A = 4.19$ rad/s^2↺; $\boldsymbol{\alpha}_B = 10.47$ rad/s^2↻.
15.19 (a) 22.7 s. (b) $\omega_A = 27.2$ rad/s↻; $\boldsymbol{\omega}_B = 45.4$ rad/s↺.
15.20 (a) $\alpha_A = 4.19$ rad/s^2↺; $\alpha_B = 6.98$ rad/s^2↺.
(b) 4.50 s.
15.21 (a) 2.22 rad/s↺. (b) 18.77 ft/s ∠40°.
15.22 (a) 2.18 rad/s↺. (b) 17.59 ft/s ∠40°.
15.23 (a) 1.25 rad/s↻. (b) 600 mm/s ⦫60°.
15.24 (a) 1.121 rad/s↻. (b) 439 mm/s ⦫60°.
15.25 (a) $\boldsymbol{\omega} = -2\mathbf{k}$ (rad/s).
(b) $\mathbf{v}_A = 4\mathbf{i} + 7\mathbf{j}$ (in./s).
15.26 $(x - 3.5\text{ in.})^2 + (y + 2\text{ in.})^2 = (3\text{ in.})^2$.
15.27 (a) $\mathbf{v}_B = 10\mathbf{i} - 3\mathbf{j}$ (in./s). (b) $x = 3.5$ in., $y = -2$ in.
15.28 (a) 2 rad/s↻.
(b) $\mathbf{v}_A = (80\text{ mm/s})\mathbf{i} + (440\text{ mm/s})\mathbf{j}$.
15.29 $(x - 220\text{ mm})^2 + (y - 80\text{ mm})^2 = (50\text{ mm})^2$.
15.30 (a) $\mathbf{v}_B = -(160\text{ mm/s})\mathbf{i} + (200\text{ mm/s})\mathbf{j}$.
(b) $x = 220$ mm, $y = 80$ mm.
15.31 (a) $\boldsymbol{\omega}_A = \boldsymbol{\omega}_B = v/r$↻; $\boldsymbol{\omega}_C = v/2r$↺.
(b) $\mathbf{v}_D = 2v$→; $\mathbf{v}_E = 0$; $\mathbf{v}_F = \sqrt{2}v$ ∠45°.
15.32 (a) 0; 84 in./s→. (b) 3.5 rad/s↺; 24.5 in./s→. (c) 0; 84 in./s←.
15.33 (a) $\mathbf{v}_P = 0$; $\boldsymbol{\omega}_{BD} = 39.3$ rad/s↺.
(b) $\mathbf{v}_P = 6.28$ m/s ↓; $\boldsymbol{\omega}_{BD} = 0$.
(c) $\mathbf{v}_P = 0$; $\boldsymbol{\omega}_{BD} = 39.3$ rad/s↻.
15.34 $\mathbf{v}_P = 6.52$ m/s ↓; $\boldsymbol{\omega}_{BD} = 20.8$ rad/s↺.
15.35 $\boldsymbol{\omega}_{BD} = 2.94$ rad/s↺; $\mathbf{v}_D = 31.8$ in./s←.
15.36 $\boldsymbol{\omega}_{BD} = 6$ rad/s↺; $\boldsymbol{\omega}_{DE} = 2$ rad/s↺.
15.37 $\boldsymbol{\omega}_{BD} = 1.2$ rad/s↺; $\boldsymbol{\omega}_{DE} = 3$ rad/s↻.
15.38 $\boldsymbol{\omega}_{BD} = 3.75$ rad/s↻; $\boldsymbol{\omega}_{DE} = 2.25$ rad/s↺.
15.39 $\boldsymbol{\omega}_{BD} = 1$ rad/s↺; $\boldsymbol{\omega}_{DE} = 3$ rad/s↺.
15.40 Vertical line intersecting zx plane at $x = 0$, $z = 5.25$ ft.
15.41 (a) 1 in. to right of A. (b) 4 in./s ↓.
(c) Outer pulley: unwrapped, 16 in./s; inner pulley: unwrapped, 8 in./s.
15.42 (a) 0.6 rad/s↻. (b) 24 mm/s→.
15.43 (a) 6.67 rad/s↻. (b) 2 m/s←.
(c) 1.250 m/s ⦪36.9°.
15.44 (a) $\boldsymbol{\omega}_{AB} = 2.25$ rad/s↺; $\boldsymbol{\omega}_{BD} = 5$ rad/s↺.
(b) 0.938 m/s ∠36.9°.
15.45 (a) 4 rad/s↺. (b) 86.5 in./s ∠16.1°.
15.46 (a) 2.31 rad/s↺. (b) 36.7 in./s ∠40.9°.
15.47 $\cos^3 \theta = b/l$.
15.48 (a) 2 rad/s↺. (b) 18.33 in./s ⦬19.1°.
15.49 (a) $\boldsymbol{\omega}_{AB} = 1.039$ rad/s↺; $\boldsymbol{\omega}_{BD} = 0.346$ rad/s↻. (b) 69.3 mm/s→.
15.50 (a) 0.25 m/s ∠36.9°. (b) 0.500 m/s ↑.
15.51 (a) 0.9 rad/s↻. (b) 144 mm/s←.

15.52 Space centrode: Horizontal rack. Body centrode: Circumference of gear.
15.53 Space centrode: Circle of 12-in. radius with center at intersection of tracks. Body centrode: Circle of 6-in. radius with center on rod at point equidistant from A and B.
15.60 5 ft/s^2 ↑ ; 0.500 rad/s^2↻.
15.61 $\mathbf{a}_A$ = 20 ft/s^2 ↑ ; $\mathbf{a}_B$ = 4 ft/s^2 ↓ .
15.62 (*a*) 0.4 m/s^2←. (*b*) 0.2 m/s^2→.
15.63 (*a*) 200 mm from B. (*b*) 225 mm from A.
15.64 (*a*) $\boldsymbol{\alpha}_A$ = 40 rad.s^2↻; $\boldsymbol{\alpha}_B$ = 20 rad/s^2↻; $\boldsymbol{\alpha}_C$ = 40 rad/s^2↻. (*b*) $\mathbf{a}_A$ = 10 in./s^2→; $\mathbf{a}_B$ = 5 in./s^2→; $\mathbf{a}_C$ = 0.
15.65 (*a*) 1.848 rad/s^2↺. (*b*) 9.24 ft/s^2 ∠60°.
15.66 (*a*) 1.848 rad/s^2↻. (*b*) 9.24 ft/s^2 ∠60°.
15.67 (*a*) 1.848 rad/s^2↺. (*b*) 8.12 ft/s^2 ∠80.2°.
15.68 (*a*) 1.848 rad/s^2↻. (*b*) 13.29 ft/s^2 ∠37.0°.
15.69 (*a*) 3.46 rad/s^2↺. (*b*) 520 mm/s^2 ∠30°.
15.70 (*a*) 3.46 rad/s^2↻. (*b*) 520 mm/s^2 ∠30°.
15.71 (*a*) 3.46 rad/s^2↺. (*b*) 454 mm/s^2 ∠7.5°.
15.72 (*a*) 3.46 rad/s^2↻. (*b*) 643 mm/s^2 ∠45.6°.
15.73 (*a*) 9.60 rad/s^2↺. (*b*) 4.05 rad/s^2↺.
15.74 (*a*) 0. (*b*) 2.67 rad/s^2↻.
15.75 $\boldsymbol{\omega}_{AB} = 0$; $\boldsymbol{\alpha}_{AB} = \frac{3}{2}\omega_0^2$↺; $\boldsymbol{\omega}_{BC} = \frac{1}{2}\omega_0$↺, $\boldsymbol{\alpha}_{BC} = 0$.
15.76 $\boldsymbol{\omega}_{AB} = \frac{1}{2}\omega_0$↺; $\boldsymbol{\alpha}_{AB} = 0$; $\boldsymbol{\omega}_{BC} = 0$; $\boldsymbol{\alpha}_{BC} = \frac{3}{2}\omega_0^2$↺.
15.77 (*a*) $\boldsymbol{\alpha}_{BC}$ = 7.5 rad/s^2↻; $\boldsymbol{\alpha}_{CD}$ = 3.75 rad/s^2↺. (*b*) 97.5 in./s^2 ∠67.4°.
15.78 (*a*) 12 rad/s^2↺. (*b*) 1.875 m/s^2 ↑ .
15.79 1.814 m/s^2 ∠60.3°.
15.80 $\omega_{AB} = r\omega(a^2 + l^2 - 2al \cos \theta)^{1/2}/al \sin \theta$.
15.81 $v_x = -3l\omega \sin \theta$, $v_y = -l\omega \cos \theta$; $a_x = -3l(\omega^2 \cos \theta + \alpha \sin \theta)$, $a_y = l(\omega^2 \sin \theta - \alpha \cos \theta)$.
15.82 $v_B = R\omega \sec^2 \theta$; $a_B = R \sec^2 \theta(\alpha + 2\omega^2 \tan \theta)$.
15.83 (*a*) $\omega = (v/R) \cos^2 \theta$. (*b*) $\alpha = -2(v/R)^2 \sin \theta \cos^3 \theta$.
15.84 (*a*) 3.09 rad/s↻. (*b*) 21.4 in./s ∠30°.
15.85 (*a*) 2.73 rad/s↻. (*b*) 0.616 m/s ∠70°.
15.86 $\boldsymbol{\omega}_{BD}$ = 3.36 rad/s↻; $\boldsymbol{\omega}_{AP}$ = 1.732 rad/s↻.
15.87 $\boldsymbol{\omega}_{BP}$ = 4.38 rad/s↺; $\boldsymbol{\omega}_{AH}$ = 1.327 rad/s↺.
15.88 (*a*) $\boldsymbol{\omega}_{BD} = \omega$↺; $\mathbf{v}_{P/AH} = 0$; $\mathbf{v}_{P/BD} = l\omega$ ↑ . (*b*) $\boldsymbol{\omega}_{BD} = \omega$↺; $\mathbf{v}_{P/AH} = 0.299\, l\omega$ ∠15°; $\mathbf{v}_{P/BD} = 1.115\, l\omega$ ∠75°.
15.89 (*a*) $\boldsymbol{\omega}_{BD} = \omega$↺; $\mathbf{v}_{P/AH} = l\omega$ ↓ ; $\mathbf{v}_{P/BD} = 0$. (*b*) $\boldsymbol{\omega}_{BD} = \omega$↺; $\mathbf{v}_{P/AH} = 0.577\, l\omega$ ∠60°; $\mathbf{v}_{P/BD} = 0.577\, l\omega$ ∠60°.
15.90 $a_P = \omega^2(r^2 + 4b^2)^{1/2}$
15.91 (*a*) 555 ft/s^2 ∠64.7°. (*b*) 690 ft/s^2 ∠46.6°.
15.92 (*a*) 0.00582 ft/s^2 west. (*b*) and (*c*) 0.00446 ft/s^2 west.
15.93 97.0 km/h.
15.94 (*a*) $\mathbf{a}_B$ = (10.9 m/s^2)**j**. (*b*) $\mathbf{a}_D$ = −(0.1 m/s^2)**i** + (10.8 m/s^2)**j**. (*c*) $\mathbf{a}_E$ = (10.7 m/s^2)**j**.
15.95 (*a*) (1 m/s^2)**i** + (10.9 m/s^2)**j**. (*b*) (2.9 m/s^2)**i** + (8.8 m/s^2)**j**. (*c*) (5 m/s^2)**i** + (10.7 m/s^2)**j**.
15.96 $a_P = R\omega^2\sqrt{1 + (2u/R\omega)^2}$.
15.97 $a_P = R\omega^2\sqrt{1 + 4/\pi^2}$.
15.98 (*a*) $\mathbf{a}_1$ = −(90.5 m/s^2)**i** − (19.99 m/s^2)**j**. (*b*) $\mathbf{a}_2$ = −(17.77 m/s^2)**i** + (57.2 m/s^2)**j**.
15.99 (*a*) 476 ft/s^2. (*b*) 307 ft/s^2.
15.100 −120 mm/s.
15.101 (*a*) $\boldsymbol{\omega}$ = (3 rad/s)**i** + (2 rad/s)**j** + (1 rad/s)**k**. (*b*) $\mathbf{v}_A$ = −(280 mm/s)**i** + (400 mm/s)**j** + (40 mm/s)**k**; $\mathbf{v}_B$ = (280 mm/s)**i** −(300 mm/s)**j** − (240 mm/s)**k**.
15.102 (*a*) $\boldsymbol{\omega}$ = (2 rad/s)**i** + (4 rad/s)**j** + (3 rad/s)**k**. (*b*) $\mathbf{v}_B$ = (5 in./s)**i** − (10 in./s)**j** + (10 in./s)**k**.
15.103 (*a*) $\boldsymbol{\omega}$ = −(1.2 rad/s)**i** + (5.4 rad/s)**k**. (*b*) $\mathbf{v}_B$ = −(27 in./s)**i** + (6 in./s)**j** − (6 in./s)**k**.
15.104 $\boldsymbol{\alpha}$ = (237 rad/s^2)**k**.
15.105 −(50.6 rad/s^2)**k**.
15.106 $\boldsymbol{\alpha}$ = −(565 rad/s^2)**i** − (5 rad/s^2)**j**.
15.107 (*a*) −(3140 rad/s^2)**k**. (*b*) +(3140 rad/s^2)**j**.
15.108 (*a*) $\boldsymbol{\omega} = -(R\omega_1/r)\mathbf{i} + \omega_1\mathbf{j}$. (*b*) $\boldsymbol{\alpha} = (R\omega_1^2/r)\mathbf{k}$.
15.109 (*a*) $-(R/r)(\omega_1 - \omega_2)\mathbf{i} + \omega_1\mathbf{j}$. (*b*) $(R/r)\omega_1(\omega_1 - \omega_2)\mathbf{k}$.
15.110 (*a*) $\boldsymbol{\alpha}$ = (3 rad/s^2)**i** + (2.5 rad/s^2)**k**. (*b*) $\mathbf{a}_A$ = −(125 in./s^2)**i** + (50 in./s^2)**j** + (67.5 in./s^2)**k**; $\mathbf{a}_B$ = −(50 in./s^2)**i** + (170 in./s^2)**j** − (180 in./s^2)**k**.
15.111 (*a*) $\boldsymbol{\omega}$ = −(4 rad/s)**j** + (1.6 rad/s)**k**. (*b*) $\boldsymbol{\alpha}$ = −(6.4 rad/s^2)**i**. (*c*) $\mathbf{v}_P$ = −(0.4 m/s)**i** + (0.693 m/s)**j** + (1.732 m/s)**k**; $\mathbf{a}_P$ = −(8.04 m/s^2)**i** − (0.64 m/s^2)**j** − (3.2 m/s^2)**k**.
15.112 $-r\omega_2^2 \cos \theta\, \mathbf{i} - r(\omega_1^2 + \omega_2^2) \sin \theta\, \mathbf{j} + 2r\omega_1\omega_2 \cos \theta\, \mathbf{k}$.

15.113 (*a*) $\boldsymbol{\alpha} = -\omega_1\omega_2\mathbf{j}$. (*b*) $\mathbf{a}_P = -r\omega_2^2\mathbf{i} + 2r\omega_1\omega_2\mathbf{k}$. (*c*) $\mathbf{a}_P = -r(\omega_1^2 + \omega_2^2)\mathbf{j}$.

15.114 (*a*) (20 rad/s)**i** − (7.5 rad/s)**j**. (*b*) (20 rad/s)**i**.

15.115 (*a*) −(2 rad/s)**i** − (4 rad/s)**j**. (*b*) −(0.4 m/s)**k**.

15.116 (*a*) $\boldsymbol{\alpha} = -(150\ \text{rad/s}^2)\mathbf{k}$. (*b*) $\mathbf{a} = -(225\ \text{in./s}^2)\mathbf{i} - (2400\ \text{in./s}^2)\mathbf{j}$.

15.117 (*a*) $\boldsymbol{\alpha} = -(8\ \text{rad/s}^2)\mathbf{k}$. (*b*) $\mathbf{a}_C = (3.2\ \text{m/s}^2)\mathbf{i} - (0.8\ \text{m/s}^2)\mathbf{j}$.

15.118 $\mathbf{v}_B = (54\ \text{mm/s})\mathbf{i}$.

15.119 +(14 mm/s)**i**.

15.120 $\mathbf{v}_C = (32\ \text{in./s})\mathbf{j}$.

15.121 (*a*) −(0.222 rad/s)**i** + (0.1906 rad/s)**j** + (0.259 rad/s)**k**. (*b*) +(54 mm/s)**i**.

15.122 (*a*) $\boldsymbol{\omega} = (1.6\ \text{rad/s})\mathbf{i} + (15.2\ \text{rad/s})\mathbf{j} - (3.2\ \text{rad/s})\mathbf{k}$. (*b*) $\mathbf{v}_C = (32\ \text{in./s})\mathbf{j}$.

15.123 (2.5 rad/s)**k**.

15.124 $\mathbf{a}_B = -(49.2\ \text{mm/s}^2)\mathbf{i}$.

15.125 $-(23.14\ \text{mm/s}^2)\mathbf{i}$.

15.126 $\mathbf{a}_C = (1162\ \text{in./s}^2)\mathbf{j}$.

15.127 $-(37.5\ \text{rad/s}^2)\mathbf{k}$.

15.128 (*a*) $-(2R\omega_1^2 + u^2/R)\mathbf{i}$. (*b*) $-R\omega_1^2\mathbf{i} - (u^2/R)\mathbf{j} + 2\omega_1 u\mathbf{k}$. (*c*) $(u^2/R)\mathbf{i}$.

15.129 (*a*) $-(u \sin 30°)\mathbf{i} - (u \cos 30°)\mathbf{j} - (r\omega_1 \cos 30°)\mathbf{k}$.
(*b*) $-\left[\left(r\omega_1^2 + \dfrac{u^2}{r}\right)\cos 30°\right]\mathbf{i} + \left(\dfrac{u^2}{r}\sin 30°\right)\mathbf{j} + (2\omega_1 u \sin 30°)\mathbf{k}$.

15.130 (*a*) $\mathbf{v}_D = -(20\ \text{in./s})\mathbf{i} - (34.6\ \text{in./s})\mathbf{j} - (46.8\ \text{in./s})\mathbf{k}$. (*b*) $\mathbf{a}_D = -(652\ \text{in./s}^2)\mathbf{i} + (133.3\ \text{in./s}^2)\mathbf{j} + (360\ \text{in./s}^2)\mathbf{k}$.

15.131 $-(702\ \text{in./s}^2)\mathbf{i} + (46.7\ \text{in./s}^2)\mathbf{j} + (256\ \text{in./s}^2)\mathbf{k}$.

15.132 (*a*) $\boldsymbol{\omega} = (0.40\ \text{rad/s})\mathbf{j} + (0.60\ \text{rad/s})\mathbf{k}$; $\alpha = (0.24\ \text{rad/s}^2)\mathbf{i}$. (*b*) $\mathbf{v}_B = (7.27\ \text{ft/s})\mathbf{i} + (4.20\ \text{ft/s})\mathbf{j} - (10.80\ \text{ft/s})\mathbf{k}$; $\mathbf{a}_B = -(6.84\ \text{ft/s}^2)\mathbf{i} + (4.36\ \text{ft/s}^2)\mathbf{j} - (5.82\ \text{ft/s}^2)\mathbf{k}$.

15.133 (*a*) $\boldsymbol{\omega} = (0.40\ \text{rad/s})\mathbf{j} + (0.60\ \text{rad/s})\mathbf{k}$; $\boldsymbol{\alpha} = (0.24\ \text{rad/s}^2)\mathbf{i}$. (*b*) $\mathbf{v}_B = (2.47\ \text{ft/s})\mathbf{i} + (16.20\ \text{ft/s})\mathbf{j} - (10.80\ \text{ft/s})\mathbf{k}$; $\mathbf{a}_B = -(14.04\ \text{ft/s}^2)\mathbf{i} + (1.483\ \text{ft/s}^2)\mathbf{j} - (1.976\ \text{ft/s}^2)\mathbf{k}$.

15.134 (*a*) $-(v_0 + R\omega_3)\mathbf{i} + \dot{R}\mathbf{j} + R\omega_1\mathbf{k}$. (*b*) $-2\dot{R}\omega_3\mathbf{i} + [\ddot{R} - R(\omega_1^2 + \omega_3^2)]\mathbf{j} + 2\dot{R}\omega_1\mathbf{k}$.

15.135 (*a*) $\mathbf{v}_P = -(1.701\ \text{m/s})\mathbf{i} + (5.95\ \text{m/s})\mathbf{j} - (3.12\ \text{m/s})\mathbf{k}$. (*b*) $\mathbf{a}_P = -(4.29\ \text{m/s}^2)\mathbf{i} - (0.201\ \text{m/s}^2)\mathbf{j} + (1.021\ \text{m/s}^2)\mathbf{k}$.

15.136 (*a*) $\boldsymbol{\omega} = \omega_1\mathbf{j} + \omega_2\mathbf{k}$; $\boldsymbol{\alpha} = \omega_1\omega_2\mathbf{i}$. (*b*) $\mathbf{v}_B = r\omega_2\mathbf{j} - (R + r)\omega_1\mathbf{k}$; $\mathbf{a}_B = -[(R + r)\omega_1^2 + r\omega_2^2]\mathbf{i}$.

15.137 $\mathbf{v}_A = r\omega_2\mathbf{i} - R\omega_1\mathbf{k}$; $\mathbf{a}_A = -R\omega_1^2\mathbf{i} + r\omega_2^2\mathbf{j} - 2r\omega_1\omega_2\mathbf{k}$.

15.138 (*a*) $\boldsymbol{\alpha} = -(0.314\ \text{rad/s}^2)\mathbf{k}$. (*b*) $\mathbf{v}_B = (124.7\ \text{ft/s})\mathbf{k}$; $\mathbf{a}_B = (25.0\ \text{ft/s}^2)\mathbf{i} - (395\ \text{ft/s}^2)\mathbf{j}$.

15.139 (*a*) $\mathbf{v}_C = (4\ \text{ft/s})\mathbf{i} - (125.7\ \text{ft/s})\mathbf{j} - (1\ \text{ft/s})\mathbf{k}$; $\mathbf{a}_C = -(0.1\ \text{ft/s}^2)\mathbf{i} - (394\ \text{ft/s}^2)\mathbf{k}$. (*b*) $\mathbf{v}_E = -(126.7\ \text{ft/s})\mathbf{k}$; $\mathbf{a}_E = -(25.2\ \text{ft/s}^2)\mathbf{i} + (395\ \text{ft/s}^2)\mathbf{j}$.

15.140 (*a*) $\boldsymbol{\alpha} = (200\ \text{rad/s}^2)\mathbf{k}$. (*b*) $\mathbf{v}_D = -(1\ \text{m/s})\mathbf{j} - (2.4\ \text{m/s})\mathbf{k}$; $\mathbf{a}_D = -(40\ \text{m/s}^2)\mathbf{i} + (44\ \text{m/s}^2)\mathbf{j} - (10\ \text{m/s}^2)\mathbf{k}$.

15.141 (*a*) $(40\ \text{m/s}^2)\mathbf{i} - (36\ \text{m/s}^2)\mathbf{j} - (10\ \text{m/s}^2)\mathbf{k}$. (*b*) $(4\ \text{m/s}^2)\mathbf{j} + (40\ \text{m/s}^2)\mathbf{k}$.

15.142 $\mathbf{v}_A = -(18\ \text{in./s})\mathbf{j} + (160\ \text{in./s})\mathbf{k}$; $\mathbf{v}_B = -(90\ \text{in./s})\mathbf{j} + (64\ \text{in./s})\mathbf{k}$; $\mathbf{a}_A = -(1600\ \text{in./s}^2)\mathbf{j} - (360\ \text{in./s}^2)\mathbf{k}$; $\mathbf{a}_B = -(880\ \text{in./s}^2)\mathbf{j} - (1000\ \text{in./s}^2)\mathbf{k}$.

15.143 $\mathbf{v}_A = -(18\ \text{in./s})\mathbf{j}$; $\mathbf{v}_B = (90\ \text{in./s})\mathbf{i} + (24\ \text{in./s})\mathbf{k}$; $\mathbf{a}_A = 0$; $\mathbf{a}_B = (480\ \text{in./s}^2)\mathbf{i} - (1000\ \text{in./s}^2)\mathbf{k}$.

CHAPTER 16

16.1 $11.72\ \text{ft/s}^2 \leftarrow$.

16.2 (*a*) 3.75 lb (*b*) **A** = 1.194 lb→; **B** = 1.194 lb←.

16.3 (*a*) 3710 N↑. (*b*) 1411 N↑.

16.4 $4.91\ \text{m/s}^2$.

16.5 5.44 m.

16.6 (*a*) 43.2 kN. (*b*) $8.38\ \text{m/s}^2 \nwarrow$.

16.7 (*a*) $11.01\ \text{ft/s}^2 \swarrow$. (*b*) 94.0 lb↖.

16.8 (*a*) $5.37\ \text{ft/s}^2 \leftarrow$ (*b*) **B** = 300 lb↑; **C** = 200 lb↑.

16.9 (*a*) $27.9\ \text{ft/s}^2 \measuredangle 60°$. (*b*) 17.83 lb.

16.10 $F_{AD} = 1.098$ lb *C*; $F_{BE} = 9.29$ lb *C*.

16.11 $\mathbf{B}_x = 205$ N←; $\mathbf{C}_x = 205$ N←.

16.12 1381 N.

16.13 $V_B = 205$ N; $M_{max} = 30.8\ \text{N} \cdot \text{m}$.

16.14 $M_B = 3.90\ \text{lb} \cdot \text{ft}$.

16.15 153.1 lb.

16.16 200 lb.
16.17 $\boldsymbol{\alpha} = \dfrac{2g}{r}\dfrac{\mu}{1+\mu}\circlearrowright$.
16.18 9.44 N.
16.19 (*a*) 19.32 rad/s$^2\circlearrowleft$. (*b*) 1.897 s.
16.20 (*a*) $\boldsymbol{\alpha}_A = 51.5$ rad/s$^2\circlearrowleft$; $\boldsymbol{\alpha}_B = 12.36$ rad/s$^2\circlearrowleft$.
(*b*) $\omega_A = 343$ rpm$\circlearrowright$; $\boldsymbol{\omega}_B = 206$ rpm$\circlearrowleft$.
16.21 (*a*) $\boldsymbol{\alpha}_A = 51.5$ rad/s$^2\circlearrowleft$; $\boldsymbol{\alpha}_B = 12.36$ rad/s$^2\circlearrowleft$.
(*b*) $\boldsymbol{\omega}_A = 1428$ rpm$\circlearrowleft$; $\boldsymbol{\omega}_B = 857$ rpm$\circlearrowright$.
16.22 (*a*) 4.91 m/s$^2\downarrow$. (*b*) 2.71 m/s$\downarrow$.
16.23 $T_A = 1582$ N; $T_B = 1385$ N.
16.24 $\mathbf{a}_A = 5.75$ m/s$^2\uparrow$; $\mathbf{a}_B = 1.747$ m/s$^2\uparrow$.
16.25 $T_A = 316$ lb; $T_B = 233$ lb.
16.26 $\mathbf{a}_A = 21.5$ ft/s$^2\uparrow$; $\mathbf{a}_B = 10.73$ ft/s$^2\uparrow$.
16.27 (*a*) 10.0 rad/s$^2\circlearrowleft$. (*b*) 1.99 m/s$^2\uparrow$.
16.28 $\boldsymbol{\alpha} = 3.72$ rad/s$^2\circlearrowright$; $\mathbf{a}_A = 19.13$ ft/s$^2 \angle 35.7°$.
16.29 (*a*) $1.098g/a\circlearrowleft$. (*b*) $1.761g \angle 36.6°$.
(*c*) $1.416g \angle 2.0°$.
16.30 (*a*) $\bar{\mathbf{a}} = \mu g\leftarrow$; $\boldsymbol{\alpha} = 2\mu g/r\circlearrowright$. (*b*) $v_0/3\mu g$.
(*c*) $5v_0^2/18\mu g$. (*d*) $\bar{\mathbf{v}} = 2v_0/3\rightarrow$;
$\boldsymbol{\omega} = 2v_0/3r\circlearrowright$.
16.31 $\dfrac{v_0^2(\mu - \tan\theta)}{2g\cos\theta\,(\frac{7}{2}\mu - \tan\theta)^2}$
16.32 75.5 mm.
16.33 $P = 4\mu W/\sqrt{58}$.
16.34 Inside a circle of radius 1 ft with center at $x = 1.5$ ft, $z = 2.5$ ft.
16.35 (*a*) 107.1 rad/s$^2\circlearrowright$. (*b*) $\mathbf{C}_x = 21.4$ N$\leftarrow$;
$\mathbf{C}_y = 39.2$ N$\uparrow$.
16.36 (*a*) 150 mm. (*b*) 125.0 rad/s$^2\circlearrowright$.
16.37 3670 lb$\rightarrow$.
16.38 (*a*) 56,200 lb. (*b*) 43,000 lb.
16.39 (*a*) $0.750g/l\circlearrowleft$. (*b*) $0.275g/l\circlearrowleft$.
16.40 (*a*) 9.81 rad/s$^2\circlearrowright$. (*b*) 442 N.
16.41 $g/2b\circlearrowright$.
16.42 (*a*) 27.8 ft/s$^2\downarrow$. (*b*) 32.2 ft/s$^2\downarrow$.
16.43 (*a*) 20.6 rad/s$^2\circlearrowright$.
(*b*) $\mathbf{A}_x = 48.3$ N$\leftarrow$, $\mathbf{A}_y = 39.3$ N$\uparrow$.
16.44 (*a*) 219 N.
(*b*) $\mathbf{E}_x = 40.1$ N$\rightarrow$, $\mathbf{E}_y = 162.9$ N$\uparrow$.
16.45 (*a*) 35.3 lb$\leftarrow$. (*b*) 2.46 lb$\rightarrow$, 26.3 lb$\uparrow$.
16.46 (*a*) 34.8 rad/s$^2\circlearrowleft$. (*b*) $\mathbf{A} = 66.6$ lb $\angle 60.9°$.
16.47 0.780 m.
16.48 1.266 m.
16.49 7.16 ft/s$^2\leftarrow$.
16.50 3.58 ft/s$^2\leftarrow$.
16.51 (*a*) $\boldsymbol{\alpha} = \frac{1}{2}P\pi/mr(\pi - 2)\circlearrowright$. (*b*) $\frac{1}{2}P/(mg + P)$.
16.52 (*a*) $\boldsymbol{\alpha} = \frac{1}{2}P\pi/mr(\pi - 2)\circlearrowright$. (*b*) $P/2mg$.
16.53 (*a*) 36.3 N$\uparrow$.
(*b*) $\mathbf{A} = 21.7$ N$\uparrow$; $\mathbf{B} = 10.80$ N$\leftarrow$.
16.54 (*a*) 23.6 N · m$\circlearrowleft$.
(*b*) $\mathbf{A} = 51.2$ N$\uparrow$; $\mathbf{B} = 6.93$ N$\leftarrow$.
16.55 (*a*) 13.23 rad/s$^2\circlearrowright$.
(*b*) $\mathbf{A} = 1.375$ lb$\uparrow$; $\mathbf{B} = 1.460$ lb $\angle 30°$.
16.56 $T_{AD} = 5.46$ lb; $T_{BE} = 6.60$ lb.
16.57 (*a*) 14.16 rad/s$^2\circlearrowright$.
(*b*) $\mathbf{A} = 17.17$ N$\uparrow$; $\mathbf{B} = 12.74$ N$\rightarrow$.
16.58 (*a*) 3.53 rad/s$^2\circlearrowright$. (*b*) 176.6 N. (*c*) 358 N$\uparrow$.
16.59 (*a*) 8.25 rad/s$^2\circlearrowleft$.
(*b*) $\mathbf{A} = 105.3$ lb$\uparrow$; $\mathbf{B} = 10.30$ lb$\rightarrow$.
16.60 $T_{BD} = 14.00$ lb; $T_{CF} = 6.36$ lb.
16.61 $\mathbf{A} = 1194$ N$\rightarrow$; $\mathbf{B} = 1876$ N$\leftarrow$.
16.62 $\mathbf{A} = 511$ N$\leftarrow$; $\mathbf{B} = 965$ N$\rightarrow$.
16.63 (*a*) 14.35 lb. (*b*) 72.1 lb.
16.64 (*a*) 4.73 lb$\uparrow$. (*b*) 5.83 lb$\uparrow$.
16.65 38.0 N$\rightarrow$.
16.66 (*a*) $\boldsymbol{\alpha}_{AB} = 3.77$ rad/s$^2\circlearrowright$;
$\boldsymbol{\alpha}_{BC} = 3.77$ rad/s$^2\circlearrowleft$.
(*b*) $\mathbf{A}_x = 15.68$ N$\rightarrow$, $\mathbf{A}_y = 43.8$ N$\uparrow$;
$\mathbf{C} = 30.2$ N$\uparrow$.
16.67 (*a*) 74.4 rad/s$^2\circlearrowright$. (*b*) 24.8 ft/s$^2\downarrow$.
16.68 (*a*) 74.4 rad/s$^2\circlearrowleft$. (*b*) 24.8 ft/s$^2\downarrow$.
16.69 (*a*) $\frac{1}{3}g\uparrow$. (*b*) $\frac{5}{3}g\downarrow$.
16.70 (*a*) $\frac{9}{7}g\downarrow$. (*b*) $\frac{3}{7}g\uparrow$.
16.71 $\boldsymbol{\alpha}_{AB} = 24.8$ rad/s$^2\circlearrowright$; $\boldsymbol{\alpha}_{BC} = 66.2$ rad/s$^2\circlearrowleft$.
16.72 $\frac{5}{16}mg\uparrow$.
16.73 $V_{max} = \frac{1}{3}mg$ at A; $M_{max} = 4mgL/81$ at $\frac{1}{3}L$ to right of A.
16.74 $\mathbf{A}_x = \frac{9}{20}W\leftarrow$, $\mathbf{A}_y = \frac{13}{40}W\uparrow$; $M_B = \frac{3}{10}WL$.
16.75 (On BC) $\frac{11}{16}M_0\circlearrowright$.

CHAPTER 17

17.1 $1.189r$.
17.2 14.06 in.
17.3 $\mathbf{v}_A = 0.964$ m/s$\downarrow$; $\mathbf{v}_B = 1.928$ m/s$\uparrow$.
17.4 $\mathbf{v}_A = 1.293$ m/s$\uparrow$; $\mathbf{v}_B = 2.59$ m/s$\downarrow$.
17.5 212 N$\uparrow$.
17.6 338 N$\uparrow$.
17.7 $v = \left[\dfrac{2mgh}{m + \dfrac{\bar{I}_A}{r_A^2} + \dfrac{\bar{I}_B}{r_B^2}}\right]^{1/2}$
17.8 9.27 ft/s$\downarrow$.

17.9 6.67.

17.10 (*a*) $l/\sqrt{12}$. (*b*) $1.861\sqrt{g/l}$.

17.11 (*a*) 6.06 ft/s. (*b*) 6.54 ft/s.

17.12 (*a*) 4.82 rad/s. (*b*) 6.81 rad/s.

17.13 $\sqrt{4gh/3}\downarrow$.

17.14 (*a*) $\bar{v} = \sqrt{\frac{10}{7}g(R - r)(1 - \cos\beta)}$.
(*b*) $N = \frac{1}{7}mg(17 - 10\cos\beta)$.

17.15 6.55 ft/s←.

17.16 4.63 ft/s←.

17.17 (*a*) $\sqrt{4gs/7}$. (*b*) $2W/7$.

17.18 (*a*) 13.45 rad/s. (*b*) 20.4 rad/s.

17.19 (*a*) $\mathbf{v}_A = 1.922$ m/s ↓;
$\mathbf{v}_B = 3.20$ m/s ↙36.9°.
(*b*) $\mathbf{v}_A = \mathbf{v}_B = 2.87$ m/s←.

17.20 (*a*) 48.2°. (*b*) 1.617 m/s.

17.21 (*a*) $\boldsymbol{\omega} = 0$, $\bar{\mathbf{v}} = \sqrt{gr}\rightarrow$. (*b*) Vertical position with end *A* at *D*.

17.22 (*a*) $\boldsymbol{\omega} = 0$, $\bar{\mathbf{v}} = \sqrt{gr}\rightarrow$. (*b*) Vertical position with end *A* at *D*.

17.23 7.67 rad/s↺.

17.24 7.13 rad/s↺.

17.25 $\mathbf{v}_B = 0$; $\mathbf{v}_D = 34.0$ ft/s ↓.

17.26 $\mathbf{v}_B = 15.03$ ft/s ↙30°; $\mathbf{v}_D = 15.03$ ft/s ↓.

17.27 (*a*) 125.7 kW. (*b*) 1257 kW.

17.28 (*a*) 0.365 lb · ft. (*b*) 1.824 lb · ft.

17.29 89.7 N · m.

17.30 439 s.

17.31 $t = \dfrac{1 + \mu^2}{1 + \mu} \cdot \dfrac{r\omega_0}{2\mu g}$

17.32 $t = (r\omega_0)/(2\mu g)$.

17.33 (*a*) Slipping occurs. (*b*) $\boldsymbol{\omega}_A = 29.9$ rad/s↻;
$\boldsymbol{\omega}_B = 39.2$ rad/s↺.

17.34 (*a*) 3.33 N · m. (*b*) $\boldsymbol{\omega}_A = 23.5$ rad/s↻;
$\boldsymbol{\omega}_B = 39.2$ rad/s↺.

17.35 (*a*) 6.58 s. (*b*) 10.83 lb; 4.17 lb. (*c*) 0.304.

17.36 (*a*) 6.58 s. (*b*) 13.06 lb; 1.945 lb. (*c*) 0.606.

17.37 (*a*) $\frac{1}{2}gt\sin\beta$↘. (*b*) $\frac{1}{2}\tan\beta$.

17.38 (*a*) 32.2 ft/s→. (*b*) Zero.

17.39 3 in.; 32.2 ft/s→.

17.40 $0.614\omega_0$.

17.41 (*a*) and (*b*) 5.71 rad/s.

17.42 (*a*) 4.38 rad/s. (*b*) 12.92 J.

17.43 (*a*) 3.09 rad/s. (*b*) 2.40 ft · lb.

17.44 (*a*) 43.5 rpm. (*b*) 1.780 ft · lb.

17.45 $v_r = 3.97$ m/s; $v_\theta = 2.86$ m/s.

17.46 (*a*) 0.522 ft. (*b*) 4.47 rad/s.

17.47 3.82 ft/s.

17.48 (*a*) Zero. (*b*) 5.23 rad/s.

17.49 (*a*) 1.333 m/s→. (*b*) 6 kN←.

17.50 (*a*) $3R/2$. (*b*) 2 m/s→.

17.51 $\mathbf{v}_A = 1.920$ ft/s←; $\mathbf{v}_B = 21.12$ ft/s→.

17.52 1.667 in.

17.53 $2r/5$.

17.54 $\boldsymbol{\omega} = \frac{1}{7}(2 + 5\cos\beta)\omega_1$↺;
$\bar{\mathbf{v}} = \frac{1}{7}(2 + 5\cos\beta)\bar{v}_1$←.

17.55 (*a*) $\bar{\mathbf{v}}_1 = mv_0/M$→; $\boldsymbol{\omega}_1 = mv_0/MR$↺.
(*b*) $mv_0/3M$→.

17.56 (*a*) $h = 3R/2$. (*b*) $h = R$.

17.57 $\boldsymbol{\omega} = 24v_1/17b$↺; $\bar{\mathbf{v}} = 12v_1/17$ ↓.

17.58 0.707.

17.59 $\dfrac{3W_S}{5W + 6W_S}\sqrt{2gh}$ ↙30°.

17.60 $h_A = h\left(\dfrac{3m}{6m + m_P}\right)^2$.

17.61 (*a*) $0.9v_0/L$↻. (*b*) $0.1v_0$→.

17.62 $\mathbf{A}\,\Delta t = m\sqrt{gl/3}$; $\mathbf{B}\,\Delta t = m\sqrt{gl/12}$.

17.63 (*a*) 50.2°. (*b*) 16.3°.

17.64 (*a*) 68.9°. (*b*) 50.2°.

17.66 $\boldsymbol{\omega}_2 = 3\bar{v}_1/2b$↻; $\mathbf{v}_x = \frac{3}{4}\bar{v}_1$→; $\mathbf{v}_y = \frac{1}{4}\bar{v}_1$ ↑.

17.67 $h_A = h\left(\dfrac{6m}{6m + m_P}\right)^2$.

CHAPTER 18

18.1 $\frac{1}{24}mL^2\omega$ ↖30°.

18.2 $\frac{1}{4}mr^2(\omega_1\mathbf{i} + 2\omega_2\mathbf{k})$.

18.3 (0.248 ft · lb · s)**i** − (0.1863 ft · lb · s)**k**.

18.4 (112.8 g · m²/s)**i** + (137.6 g · m²/s)**j**
− (144 g · m²/s)**k**.

18.5 −(0.667 kg · m²/s)**i** + (0.667 kg · m²/s)**j**.

18.6 (0.222 kg · m²/s)**i** + (0.222 kg · m²/s)**j**.

18.7 (2.72 ft · lb · s)**i** + (2.70 ft · lb · s)**j** +
(2.88 ft · lb · s)**k**.

18.8 (9.37 ft · lb · s)**i** + (7.32 ft · lb · s)**j** −
(10.50 ft · lb · s)**k**.

18.9 (*a*) −(6.36 kg · m²/s)**i** + (2.12 kg · m²/s)**j** +
(14.14 kg · m²/s)**k**. (*b*) 154.6°.

18.10 (*a*) $m\bar{\mathbf{v}}$ = (240 Mg · m/s)**i** + (360 Mg · m/s)**j**
+ (72 × 10³ Mg · m/s)**k**;
$\mathbf{H}_G$ = (432 Mg · m²/s)**j** + (180 Mg · m²/s)**k**.
(*b*) 67.1°.

18.11 (*a*) $-(F\,\Delta t/m)\mathbf{k}$. (*b*) $(12F\,\Delta t/7ma)(-\mathbf{i} - 5\mathbf{j})$.

18.12 (*a*) $-(F\,\Delta t/m)\mathbf{k}$. (*b*) $(12F\,\Delta t/7ma)(2\mathbf{i} + 3\mathbf{j})$.

18.13 (*a*) $(6F\ \Delta t/7ma)(\mathbf{i} - 7\mathbf{j})$. (*b*) Axis through *A*, in *xy* plane, forming ∢81.9° with *x* axis.
18.14 (*a*) $\frac{5}{6}\omega_0(\mathbf{i} - \mathbf{j})$. (*b*) $-\frac{5}{6}\omega_0 a\mathbf{k}$.
18.15 (0.400 rad/s)**i** − (0.300 rad/s)**j** + (0.075 rad/s)**k**.
18.16 −(6.18 rpm)**i** − (2.86 rpm)**j** + (0.716 rpm)**k**.
18.17 (*a*) $\Delta t_A = 1.213$ s; $\Delta t_B = 0.558$ s. (*b*) $\Delta\bar{\mathbf{v}} = (0.0886\text{ m/s})\mathbf{k}$.
18.18 (*a*) 0.320 s. (*b*) (0.0795 rad/s)**j**. (*c*) (0.0160 m/s)**k**.
18.19 (5.97 rpm)**i** − (2.69 rpm)**j** + (0.806 rpm)**k**.
18.20 −(6.03 rpm)**i** − (2.69 rpm)**j** + (0.806 rpm)**k**.
18.21 0.497 ft · lb.
18.22 $-(5/48)ma^2\omega_0^2$.
18.23 (*a*) (42.4 rpm)**j** + (64.2 rpm)**k**. (*b*) 4050 J.
18.24 −2.06 ft · lb.
18.25 $-\frac{1}{2}mr^2\omega_1\omega_2\mathbf{j}$.
18.26 (0.800 N · m)**k**.
18.27 (0.745 lb · ft)**j**.
18.28 (0.497 lb · ft)**i** + (0.745 lb · ft)**j** − (0.373 lb · ft)**k**.
18.29 **A** = (46.2 N)**j**; **D** = −(46.2 N)**j**.
18.30 $\mathbf{A} = \frac{1}{4}\sqrt{2}ma\omega^2\mathbf{k}$, $\mathbf{M}_A = \frac{1}{6}ma^2\omega^2\mathbf{i}$.
18.31 $\mathbf{A} = \frac{1}{2}(w/g)a^2\omega^2\mathbf{k}$; **B** = −**A**.
18.32 **A** = (2.80 lb)**i**; **B** = −(2.80 lb)**i**.
18.33 (*a*) (44.4 rad/s²)**k**. (*b*) **A** = −**B** = −(19.17 N)**i** + (17.50 N)**j**.
18.34 (*a*) $(4M_0/ma^2)\mathbf{j}$. (*b*) $\mathbf{R}_A = -(M_0\sqrt{2}/a)\mathbf{i}$; $\mathbf{M}_A = \frac{2}{3}M_0\mathbf{k}$.
18.35 (*a*) **M** = (0.647 lb · ft)**i**. (*b*) **A** = (0.388 lb)**j**; **B** = −(0.388 lb)**j**.
18.36 (*a*) (28.0 lb · ft)**j**. (*b*) **A** = (2.33 lb)**k**; **B** = −(2.33 lb)**k**.
18.37 −(0.831 N · m)**i**.
18.38 7.07 kN · m.
18.39 22.7 lb · ft.
18.40 A = 1.16 lb ↑; B = 4.84 lb ↑.
18.41 (*a*) $2mr^3\omega^2$. (*b*) 0. (*c*) $\frac{1}{2}mr^3\omega^2$.
18.42 (*a*) $\cos\beta = \dfrac{gl}{\omega^2(l^2 - \frac{1}{4}r^2)}$. (*b*) $\omega = \sqrt{gl/(l^2 - \frac{1}{4}r^2)}$.
18.43 $\frac{1}{2}mr^2\omega_1\omega_2\mathbf{i}$.
18.44 $\mathbf{F} = -mR\omega_1^2\mathbf{i}$; $\mathbf{M}_0 = \frac{1}{2}mr^2\omega_1\omega_2\mathbf{i} - mRh\omega_1^2\mathbf{k}$.
18.45 (*a*) $\sqrt{g/a}$. (*b*) $\sqrt{2g/a}$.
18.46 21.1 lb ↓.
18.47 **D** = −(0.622 N)**j** − (4.00 N)**k**. **E** = (3.82 N)**j** − (4.00 N)**k**.
18.48 (*a*) $M_1 = \frac{1}{24}mL^2\omega_2^2 \sin 2\theta$. (*b*) $M_2 = -\frac{1}{12}mL^2\omega_1\omega_2 \sin 2\theta$.
18.49 4450 rpm.
18.50 3500 rpm.
18.51 (*a*) 2.75 rpm. (*b*) 2.77 rpm; 397 rpm.
18.52 Precession axis: ⦣30°; precession, 6.00 rad/s; spin, 10.39 rad/s.
18.53 Precession axis: $\theta_x = 39.9°$, $\theta_y = 127.9°$, $\theta_z = 79.4°$; precession, 4.38 rpm; spin, 2.61 rpm.
18.54 Precession axis: $\theta_x = 140.4°$, $\theta_y = 127.6°$, $\theta_z = 79.5°$; precession, 4.39 rpm; spin, 2.65 rpm.
18.55 Precession axis: $\theta_x = 90°$, $\theta_y = 58.0°$, $\theta_z = 32.0°$; precession, 1.126 rpm (retrograde); spin, 0.343 rpm.
18.56 Precession axis: $\theta_x = 90°$, $\theta_y = 38.0°$, $\theta_z = 52.0°$; precession, 1.454 rpm (retrograde); spin, 0.322 rpm.
18.57 $3\sqrt{g/2L}$.
18.58 (*a*) $(1 + \cos^2\theta)\dot{\phi}^2 + \dot{\theta}^2 = \text{constant}$; $(1 + \cos^2\theta)\dot{\phi} = \text{constant}$. (*b*) $\dot{\theta} = \dot{\phi}_0\sqrt{\dfrac{(1 + \cos^2\theta_0)(\cos^2\theta - \cos^2\theta_0)}{1 + \cos^2\theta}}$
18.59 $\dot{\phi} = 4\dot{\psi}_0/15$; $\dot{\psi} = 17\dot{\psi}_0/15$.
18.60 (*a*) $5\sqrt{3g/2a}$. (*b*) $\dot{\phi} = \sqrt{3g/2a}$; $\dot{\psi} = 5\sqrt{3g/2a}$.

CHAPTER 19

19.1 (*a*) 0.1900 m. (*b*) 2.39 m/s.
19.2 0.257 s; 1.837 m/s; 45.0 m/s².
19.3 3.49 ft/s; 36.5 ft/s².
19.4 (*a*) 0.506 s. (*b*) 2.07 ft/s. (*c*) 25.8 ft/s².
19.5 1.585 in. ↑; 1.263 ft/s ↑; 20.4 ft/s² ↓.
19.6 (*a*) 0.1685 s. (*b*) 1.793 ft/s ↑; 12.88 ft/s² ↓.
19.7 (*a*) 0.294 s; 3.40 Hz. (*b*) 0.428 m/s; 9.14 m/s².
19.8 (*a*) 0.679 s; 1.473 Hz. (*b*) 0.1852 m/s; 1.714 m/s².
19.9 *Pendulum:* $\tau = 2\pi\sqrt{l/(g + a)}$; *Mass & Spring:* $\tau = 2\pi\sqrt{m/k}$.
19.10 (*a*) 0.994 m. (*b*) 3.67°.

19.11 4.55 lb.
19.12 (*a*) 4.53 lb. (*b*) 0.583 s.
19.13 (*a*) 1.876 s. (*b*) 7.07°.
19.14 393 mm.
19.15 1.566 s.
19.16 0.971 s.
19.17 (*a*) 0.456 s. (*b*) 0.689 m/s.
19.18 (*a*) 0.302 s. (*b*) 1.042 m/s.
19.19 (*a*) 0.923 s. (*b*) 1.135 ft/s.
19.20 (*a*) 1.084 s^{-1}.
(*b*) $T_B = 16$ lb; $T_C = 12.40$ lb.
19.21 $\tau = 2\pi\sqrt{2m/3k}$.
19.22 $f = (1/2\pi)\sqrt{2k/m}$.
19.23 (*a*) $f = (1/2\pi)\sqrt{4k/m}$.
(*b*) $f = (1/2\pi)\sqrt{12k/m}$.
19.24 $f = (1/2\pi)\sqrt{3k/m}$.
19.25 (*a*) $\tau = 2\pi\sqrt{7l/6g}$. (*b*) $\tau = 2\pi\sqrt{5l/6g}$.
19.26 (*a*) $f = (1/2\pi)\sqrt{g/2r}$. (*b*) $f = (1/2\pi)\sqrt{2g/3r}$.
19.27 $\bar{k}_x = 6.08$ ft; $\bar{k}_z = 6.74$ ft.
19.28 8.60 ft.
19.29 (*a*) 5.44 s. (*b*) 1.089 m/s.
19.31 (*a*) 8.13 s. (*b*) 1.820 ft/s.
19.32 0.456 in.
19.37 $\tau = 6.82\sqrt{l/g}$.
19.38 $\tau = 2\pi\sqrt{0.866l/g}$.
19.39 $\tau = 2\pi\sqrt{0.866l/g}$.
19.41 $f = (1/2\pi)\sqrt{g/2l}$.
19.42 0.926.
19.43 9.90 s.
19.46 $\tau = 2\pi\sqrt{m/3k\cos^2\beta}$.
19.47 $\tau = 2\pi\sqrt{(\frac{1}{3}m + m_c)/k\cos^2\beta}$.
19.48 $\tau = 2\pi\sqrt{m_c/k\cos^2\beta}$.
19.49 0.703 mm.
19.50 (*a*) 11.38 μm. (*b*) 320 μm. (*c*) ∞.
19.51 (*a*) 400 rpm. (*b*) 0.00167 in.
19.52 (*a*) 168.0 rpm. (*b*) 0.00131 in.
19.53 1007 rpm.
19.54 0.794 mm.
19.55 0.750 in. or 0.1875 in.
19.56 0.0857 in. or 0.120 in.
19.57 1085 rpm.
19.58 122.5 rpm.
19.59 109.5 rpm and 141.4 rpm.
19.60 0.583 Hz; 0.452 Hz.
19.61 1.200 mm.
19.62 (*a*) 40.5 km/h. (*b*) 25.1 mm.
19.63 70.1 km/h.
19.64 72.5 μm.
19.65 (*a*) 0.0355. (*b*) 8.98 N · s/m.
19.66 1320 lb · s/ft.
19.67 (*a*) $x = x_0e^{-pt}(1 + pt)$. (*b*) 0.1108 s.
19.68 7780 lb/ft.
19.69 (*a*) 1.509 mm. (*b*) 0.583 mm.
19.70 (*a*) 0.1514. (*b*) 313 N · s/m.
19.71 0.1791 in.
19.72 10.61 lb · s/in.
19.73 (*a*) 270 rpm. (*b*) 234 rpm. (*c*) 8.84 mm; 9.45 mm.
19.74 $m\ddot{x}_A + 5kx_A - 2kx_B = 0$;
$m\ddot{x}_B - 2kx_A + 2kx_B = P_m \sin \omega t$.
19.77 (*a*) $m_1\ddot{x}_1 + c_1\dot{x}_1 + k_1x_1 + k_2(x_1 - x_2) = 0$,
$m_2\ddot{x}_2 + c_2\dot{x}_2 + k_2(x_2 - x_1) = 0$.
(*b*) $L_1\ddot{q}_1 + R_1\dot{q}_1 + q_1/C_1 + (q_1 - q_2)/C_2 = 0$, $L_2\ddot{q}_2 + R_2\dot{q}_2 + (q_2 - q_1)/C_2 = 0$.
19.78 (*a*) $m_1\ddot{x}_1 + c_1\dot{x}_1 + c_2(\dot{x}_1 - \dot{x}_2) + k_1x_1 + k_2(x_1 - x_2) = 0$; $m_2\ddot{x}_2 + c_2(\dot{x}_2 - \dot{x}_1) + c_3\dot{x}_2 + k_2(x_2 - x_1) + k_3x_2 = P_m \sin \omega t$.
(*b*) $L_1\ddot{q}_1 + R_1\dot{q}_1 + R_2(\dot{q}_1 - \dot{q}_2) + \dfrac{q_1}{C_1} + \dfrac{q_1 - q_2}{C_2} = 0$; $L_2\ddot{q}_2 + R_2(\dot{q}_2 - \dot{q}_1) + R_3\dot{q}_2 + \dfrac{(q_2 - q_1)}{C_2} + \dfrac{q_2}{C_3} = E_m \sin \omega t$.